Phytochemistry and Pharmacology of Garlic (*Allium sativum* L.) Clove

The present book is an attempt to provide a thorough overview on the origin of garlic, history of garlic chemistry, garlic preparations and their use in folklore, health benefits of garlic, garlic as pesticide and its adverse effects. Garlic (*Allium sativum*) is also used as a popular herbal alternative therapy. There is a lack of consolidated literature on the health benefits of garlic, this book will provide a comprehensive compendium on these benefits. Though it is a scientific research-based knowledge compilation, it will be also equally beneficial to science students as well as the lay reader.

Salient features:

- Focuses on the origin and history of garlic, its botanical characterization and preparation
- Discusses health benefits of garlic based on scientific studies, research and their applications
- Deals with pesticidal, insecticidal, molluscicidal and nematicidal activities of garlic
- Elucidates the benefits of garlic in cardiovascular diseases, antidiabetic effect, anticancer potential and antioxidant activity

Phytochemistry and Pharmacology of Garlic (*Allium sativum* L.) Clove

Dinesh Kumar Singh and Devesh Pratap Singh

CRC Press
Taylor & Francis Group
Boca Raton London New York

CRC Press is an imprint of the
Taylor & Francis Group, an **informa** business

Designed cover image: ShutterStock image id: 1458628313

First edition published 2026
by CRC Press
2385 NW Executive Center Drive, Suite 320, Boca Raton FL 33431

and by CRC Press
4 Park Square, Milton Park, Abingdon, Oxon, OX14 4RN

CRC Press is an imprint of Taylor & Francis Group, LLC

ISBN: 978-1-041-08606-2 (hbk)
ISBN: 978-1-041-08607-9 (pbk)
ISBN: 978-1-003-64621-1 (ebk)

DOI: 10.1201/9781003646211

Typeset in Times
by Apex CoVantage, LLC

We dedicate this book to our research mentor and pathfinder of our academic journey

Lt. Prof. R. A. Agarwal, Ex-Head, Department of Zoology, Ex-Dean Faculty of Science,

Ex-Pro Vice-Chancellor, D.D.U. Gorakhpur University, Gorakhpur, U.P., India

Contents

Preface

Nowadays, the popularity of using herbs and other alternative therapies is rising significantly. Among them, the use of garlic is one of the most popular herbal alternative therapy. There are over 300 strains of garlic grown all over the word. Garlic is mentioned in the Bible, literature of ancient Israel, Babylonia, Greek, Rome, China and India. In ancient times, around 3000 BC, the father of Ayurveda, Charaka, used garlic to maintain the fluidity of blood and strengthen the heart. In the Talmud, an important collection of ancient rabbinic writings, garlic is recommended for treating wounds. Roman naturalist, Pliny the Elder, who is the author of the first encyclopedia, *Historia Naturalist*, wrote that garlic could cure 62 ailments, including hemorrhoids, dog and snake bites and tumors. Hippocrates recommended garlic as a laxative and diuretic. Aristotle recommended it as a cure for rabies. In 1550 BC, an Egyptian papyrus, the *Codex Ebers*, give 22 garlic-based therapeutic formulas for different ailments. Prophet Mohammad recommended it for treating scorpion stings.

In ancient Egypt, garlic had the reputation of giving stamina and physical strength. It is told that workers engaged in building the Great Pyramid received a daily ration of garlics and onions. During the construction of one pyramid, a garlic shortage forced the Egyptians to cut slaves' rations. The result was the world's first recorded strike. In the year 1721, during a horrible plague epidemic in Europe, four condemned criminals were forced to burry the dead bodies. They appeared to be immune to the disease; this was attributed to their drinking a preparation consisting of macerated garlic in wine. The concoction became famous as "Vinaigre des quatre voleurs". Even today, it is available in France. Louis Pasteur in the 1800s had reported that garlic kills bacteria.

The Chinese believe that garlic imparted longevity. An ancient Chinese sage is reported to have said, "When you are 50 years old, then you should eat garlic for 50 days, and you will live for a further 50 years". Tremendous valuable benefits have been reported in various historical writings. Garlic is used as a nutritive element in food. It is used in the treatment of rheumatism, dermatitis, abdominal pain, cough, loss of appetite, tumors, cure of rabies, dog and snake bite, fever, asthma, epilepsy, vermifuge, etc. The use of garlic decreases the aging process and prolongs life.

Prof. Dinesh Kumar Singh
Dr. Devesh Pratap Singh

Acknowledgements

With a profound sense of gratitude, I am thankful to Lt. Prof. K.C. Srivastava, Department of Environmental Medicine, Institute of Community Health, Odence University, Odence, Denmark, who has given his valuable suggestions on garlic throughout my research journey. He was a world-renound scientist in the field of spices. Prof. Srivastava had studied the pharmacological effects of spices on human health. He introduced me to various scientists who are working on spices. Prof. Erick Block is one of them. I am also thankful to Prof. Erick Block, Department of Chemestry, University of Albany, State University of New York, Albany, New York, USA. Prof. Erick Block has given sufficient support through his valuable research publications.

We take this opportunity to acknowledge all the members of Malacology Laboratory, Department of Zoology, D.D.U. Gorakhpur University, Gorakhpur—273009, U.P., India.

The author is greatful to Mr. Ajay Pati Tripathi for his expert editorial helps. Last but not the least, we wish to thank our families and friends for their continuous support.

Prof. Dinesh Kumar Singh
Dr. Devesh Pratap Singh

About the Authors

Prof. Dinesh Kumar Singh, Ex-Head, Department of Zoology, DDU Gorakhpur University, Gorakhpur, U.P., India, is actively engaged in the study of molluscan physiology, biochemistry and snail control for four decades. He has successfully isolated and identified pure molluscicidal compounds from more than three dozen medicinal plants and observed their mode of action on snails at the cellular level. His work is recognized internationally in the field of formulation and use of synthetic and biopesticides for integrated vector management. Prof. DK Singh has the honor to act as the referee for the Third World Academy award sponsored by TWA, Italy. He is recognized as a member of the international fasciolosis control team of WHO. His expertise for controlling snails has been sought by scientific bodies in the Philippines and African countries. Prof. Singh has published 217 full research papers and 17 review articles in 112 journals of international repute. Out of 217 papers, he has published 22 full papers on garlic. He has written five reference books and three book chapters. Recently, in 2021, he published a book entitled *Fasciolosis: Causes, Challenges and Controls* with Springer Nature and a chapter, "Phytochemestry of *Sapindus mukrossi* and Medicinal Properties", in *Medicinal and Aromatic Plants of the World* in the *Encyclopedia of Life Support Systems* (EOLSS), developed under the auspices of the UNESCO-EOLSS Publishers, Paris, France. He has produced 27 Ph.D. graduates and successfully conducted 17 research projects funded by various scientific agencies: UGC, CSIR, DBT, UPCST, DST and ICAR, India. Prof. Singh is a fellow of the Zoological Society of India and the Academy of Environmental Biology. He is a life member of national and international academic societies and an editor and reviewer of various international journals. His research team reported, for the first time, a nucleotide sequencing of *Fasciola gigantica* and their larva cercaria (GenBank accession number KT149356 and KT149357 for cercaria larva and *F. gigantica*, respectively. Source-www.ncbi.nlm.nih.gov/KT149357.1 and /KT149357.2).

Dr. Devesh Pratap Singh is working as Associate Professor of Pulmonary, Critical Care and Sleep Medicine at All India Institute of Medical Sciences, Gorakhpur, U.P., India. He has published 25 papers in journals of international repute and has ten years of experience in medical teaching, as well as in interventional pulmonology.

About the Authors

About the Book

Garlic (*Allium sativum*) is used as a popular herbal alternative therapy. There is a lack of consolidated literature on health benefits of garlic. The present book is an attempt to make a comprehensive overview on the origin of garlic, history of garlic chemistry, garlic preperations and their use in folklore, health benefits of garlic, garlic as pesticide and its adverse effects. The first chapter of the book focuses on the origin and history of garlic, its botanical characterization and preperations. The second chapter contains the chemistry of the garlic clove and its different preperations, whereas the third chapter covers garlic's uses in folklore. The fourth chapter is on the health benefits of garlic based on scientific studies, research and their applications. The fifth chapter deals with the pesticidal/insecticidal/molluscicidal/nematicidal activities of garlic. The sixth chapter deals with the adverse effects of garlic. The historical dimensions of garlic use in ancient times as well as in the modern scientific world are discussed in this book. The chemistry of the garlic clove, its different preperations and their uses for different health benefits are described. The health benefits of garlic—antimicrobial (antibacterial, antifungal, antiviral, antiprotozonal) activities, effects on cardiovascular diseases, antidiabetic effects, anticancer potential, antioxidant activity, neuroprotection, use as a pesticide and adverse effects—are given based on scientific studies and research. The author is confident that this book is a step towards a new approach to explore the health benefits of garlic.

1 Overview on Garlic

1.1 RECENT TRENDS IN GARLIC STUDIES

Garlic (*Allium sativum* L.) has been considered as a valuable health-improving agent by people of different cultures for thousands of years. It is commonly used in Eastern Europe and Asia for its medicinal benefits. In Germany, sales of garlic preperations rank those of the leading prescription drugs. Garlic is also one of the most researched medicinal plants. From 2002 to 2016, 2,836 papers have been published on garlic research. This study indicates that year-wise publications on garlic research have increased from 4.76% to 7.48% (Sivasami, 2017). Relative growth rate means the value of garlic research publication and doubling time was 0.5 and 10.41. Out of 2,836 research publications, articles, abstracts and reviews were 2,154 (75.95%), 343 (12.09%) and 83 (2.93%), respectively. Among the 92 countries that published 2,836 research papers on garlic, the top five are as follows: the United States of America (USA), which occupies first place with 16.26% of the papers; South Korea, in second place with 11.11%; India, in occupies third place with 9.45%; China, in fourth place with 9.31%; and Iran, in fifth place with 5.75%.

Among 2,228 institutions, the top five that contributed to garlic research over the past 15 years are as follows: the Council of Scientific Industrial Research (CSIR), India, occupies first place with 52 papers; the United States Departmental of Agriculture (USDA) is in second place with 51 papers; the Univesity of California is in third place with 50 papers; Wakunaga Pharmaceut Co. Ltd. is in fourth place with 36 papers; and Seoul National University is in fifth place with 35 papers. In 15 years (2002–2016), a total 8,130 authors have contributed to garlic research. Among them, the top five authors are as follows: Ide, N. holds first place with 27 papers; Change, ZH is in second place with 24 contributions; Kyung, KH is in third place with 22 contributions; Ali, M. is in fourth place with 21 papers; and Tomson, M. is in fifth place with 19 papers (Sivasami, 2017).

Among the 2,836 papers, those with six or more authors accounted for 27.26%; five-author papers made up 14.07%; four-author papers, 19.08%; three-author papers, 18.34%; two-author papers, 14.42%; and single-author papers, 6.84%. Among the 2,836 papers published in 15 years, 93.16% of authors are collaborative authors (Sivasami, 2017). Scientometric analysis also demonstrated that 991 sources contributed to the publication of garlic papers. Among them, the top five sources are as follows: the *Journal of Agricultural Food Chemistry* occupies the first position with 70 papers; the *Journal of Nutrition* is in second place with 65 papers; the *FASEB Journal* holds third place with 64 papers; *Plant Disease* is in fourth place with 43 publications; and *Food Science and Biotechnology* is in fifth place with 36 papers (Sivasami, 2017).

Fenwick and Hanley (1985a, 1985b), Hanley and Fenwick (1985), Sprecher (1986), Reuter (1995), Abdullah et al. (1988), Petkov (1966) and Koch and Lawson (1996)

DOI: 10.1201/9781003646211-1

reviewed the hormonal action and metabolism of garlic. The most extensive review on the biological and chemical studies on garlic was published by Koch and Hahn (1988). Block (1992) published review articles on the chemistry of garlic.

Reviews on the chemistry of garlic have been written by Whitaker (1976), Raghavan et al. (1983), Fenwick and Hanley (1985a, 1985b), Hanley and Fenwick (1985), Block (1985, 1986), Lutomski (1987), Lancaster and Boland (1990) and Lawson (1993). Extensive reviews on the effects of garlic have been written on cardiovascular disease (Kleijnen et al., 1989; Lau et al., 1983; Turner, 1990; Kendler, 1987; Ernst, 1987) and cancer (Lau et al., 1990; Sumiyoshi and Wargovich, 1989). In 1993, Srivastava and Mustafa (1993) and Srivastava et al. (1995) wrote extensive review articles on the pharmacology and disease prevention benefits of garlic. Eric Block (1992) discussed the organosulfur chemistry of the genus *Allium* in his review article. Morbidoni et al. (2001) wrote the history and adverse effects of garlic in their review article "Garlic: Its history and adverse effects". Harris et al. (2001), in their review article, discussed the antimicrobial properties of garlic. In 2008, Singh and Singh wrote an extensive review article on the pharmacology of *Allium sativum*. Professor Eric Block (2010), an unrivaled authority on every aspect of garlic and *Alliums*—from cultural, historical, botanical, environmental, culinary and medical effects to their fundamental science—provides an extensive overview of the chemistry of garlic in his book *Garlic and Other Alliums*. Alam et al. (2016) wrote a review article on the medicinal properties of garlic. Singh et al. (2020) wrote a review on the pharmaceutical beneficial/adverse uses of garlic for human health. Batiha et al. (2020) discussed the chemical constituents and pharmacological activities of garlic in a review article published in *Nutrients*. Tellez et al. (2020) wrote a review on the beneficial properties of garlic for health. Tesfaye and Mengesha (2015) and Tesfaye (2021) wrote a review article on the therapeutic uses of garlic and its role in drug discovery. Tadu et al. (2022) provided the traditional uses, phytochemistry, pharmacology and toxicology of garlic in their review article published in *Frontiers in Nutrition*. Recently, Bazaraliyeva et al. (2022) provided the chemical and biological properties of the bioactive compounds of garlic.

Garlic is traditionally eaten as the cooked or chopped clove. Nowadays, it is consumed in the form of encapsulated commercial products. When used for health reasons, clinical studies on garlic have employed a processed garlic product. Several forms of garlic products, frequently called garlic in both lay and scientific literature, differ in composition from whole or crushed garlic cloves. The chemistry of garlic products varies qualitatively and quantitatively according to the processing method utilized and the source (Lawson, 1993; Block, 2010).

1.2 ORIGIN OF GARLIC

Common garlic, *Allium sativum* L., appears to have originated in Central Asia and then spread into China, the Near East and the Mediterranean regions before moving west to Central and Southern Europe, Northern Africa and Mexico (Singh and Singh, 2008). Some historians believe garlic was native to Siberia and was spread to other geographic regions more than 5,000 years ago by Siberian nomadic tribes (Morbidoni et al., 2001).

There are several stories of folklore to explain the garlic's beginning. A story from the 5th century AD Bower Manuscript says that garlic grew from the drops of blood that fell to the earth when Janardana decapitated Asura, the ruler of the old Indian gods (Koch and Lawson, 1996). According to the ancient Islamic religious leader Mohammad, garlic sprang from the left footprint and onion from the right footprint of Satan as he walked out of the Garden of Eden (Day, 1999).

According to recorded historical evidence, garlic has played rich and diverse commercial, culinary, literary and mythical roles. Ancient Egyptian and Middle Eastern cultures used garlic as a currency. Around 2500 BC, 15 pounds of garlic could purchase a healthy slave (Christopher Ranch, 1999; Morbidoni et al., 2001). In Southern Europe, the Balkans rubbed garlic juice on doorknobs and window frames to discourage vampires from haunting them. They hung garlic on doors so that nobody would take milk from their cows and the families would be safe from witches (Koch and Lawson, 1996; Christopher, 1999). Whereas in other parts of Europe, fishermen and seamen used to wear cloves of garlic to protect themselves from evil spirits and diseases (Koch and Lawson, 1996).

In various religions and cultures, shunning garlic use is believed to be evil. Certain sects of Christianity, Hinduism, Islam, Jainism and Buddhism have considered garlic unclean (Fenwick and Hanley, 1985a, 1985b; Day, 1999). In ancient Greece, people who had eaten garlic were forbidden from entering the temple of Cybele because its smell was considered affensive (Day, 1999).

1.3 HISTORICAL PROSPECTIVE OF GARLIC USE

The teachings of the father of ayurvedic medicine, Charaka (3000 BC), and the Egyptian medical manuscript *Codex Ebers* (3,500 year ago) mentioned the uses of garlic to maintain the fluidity of blood and strengthen the heart (Fenwick and Hanley, 1985a, 1985b; Kendler, 1987; Farbman et al., 1993). In the ancient rabbinic writing Talmud, garlic is mentioned for treating wounds (Farbman et al., 1993). According to Pliny the Elder, who is the author of the first encyclopedia, *Historia Naturalis*, garlic could be used to cure 62 ailments (Fenwick and Hanley, 1985a, 1985b). Hippocrates, the father of medicine, recommended garlic as a laxative and diuretic (Koch and Lawson, 1996; Fenwick and Hanley, 1985a).

The use of garlic was known to the Chinese since time immemorial. They believed that the use of regular garlic could enhance the longevity of life. They strongly believed that garlic is useful in most human diseases. Garlic can be useful against cancer, heart disease and various kinds of infections. The Babylonians used garlic against leprosy, cholera and typhoid. During the outbreak of the plague in London in the 18th century, French priests who attended the plague victims remained healthy, whereas the English priests succumbed. This difference is due to the use of garlic; the French priests consumed it, while their English counterparts detested it (Roser, 1991).

Adam Lonitzer, a German physian, author of a herbal book, mentioned that garlic juice can be used externally for killing lice and nits and internally to rid of the body from worms and poison (Koch and Lawson, 1996). Louis Pasteur in 1800 has reported that garlic can kill bacteria (Farbman et al., 1993). In the Second World War, garlic juice with water was used to disinfect wounds (Bolton et al., 1983).

1.4 BOTANICAL CHARACTERIZATION

Garlic belongs to the family Amaryllidaceae. It is widely grown in different parts of the world. Garlic is a perennial plant with a poorly developed root system, consisting of 40 to 100 secondary roots that grow out of a shortened. Lignified stalk heel is 3 to 8 cm high and half-leaved. Leaves of garlic are divided into protective scales enfolding the bulb, storage leaves for following food reserves and photoassimilating leaves for carbohydrates production. The photoassimilating leaves are flat, rough at the edges, approximately 1 cm wide, ensiform and narrowing.

The two main subspecies of garlic are hardneck and softneck. Hardneck garlic (*Allium sativum* spp. Ophioscorodon) produces scapes or flower stalk and prefers cold winters. It contains a single circle of 6 to 11 cloves around a central woody stalk. Before flowering, the top set curls upward as it grows, forming one to three coils before straightening and then producing little seedlike bulbils. Softneck garlic (*Allium sativum* ssp. Sativum) contains 24 cloves per bulb in several layers around a soft central stem. Large cloves are outside and smaller cloves are in the middle. It only develops a short scape lacking a flower top, so it is also called a short-necked garlic. Softneck garlic usually has better storage ability than hardneck garlic and is preferred for braiding since its neck remains soft at harvest time (Block, 2010). On the basis of DNA analysis, agriculture geneticist Gayle Volk classifies the garlic into ten major types. They are rocambole, purple stripe, marble purple stripe, glazed purple stripe, porcelain, artichoke, silverskin, asiatic, turban and creole. Artichoke and silverskin are major softneck garlic types. Silverskin garlic is commonly found in grocery stores due to its long storage life (Volk et al., 2004; Volk and Stern, 2009). Garlic is completely sterile and is, therefore, propagated asexually only from cloves (Shemesh et al., 2008).

Fertile wild garlic has been discovered and collected at its center of origin in Kazakhstan and Kyrgyzstan. The wild ancertons of garlic are the only source of germplasm for the development of new cultivars. Many of the older varities of garlic demonstrate superior resistance to pests and diseases (Kamenetsky et al., 2005).

Local genotypes from Central Asia have wide genetic diversity, including economically useful traits lost during the 10,000 years of garlic selection by man (Block, 2010). The use of garlic seed for breeding purposes would eliminate the main ailments associated with the clonal propagation, which carry over pests from one generation to another, low propagation rates, voluminous storage of bulbs, rotting and sprouting. The use of seeds would save the costs of vegetative propagation and help in virus elimination (Shemesh et al., 2008). Although the garlic genome has been mapped and garlic seed production is now routine (Ipek et al., 2005), wide-scale commercial production of garlic from seeds is not performed, as seed-grown garlic is neither competitive in appearance nor cost with the garlic of China (Block, 2010).

1.5 GARLIC PREPARATIONS

Garlic is a food product composed of various macronutrients, vitamins, trace elements, organosulfur compounds and multiple pharmacologically active ingradients. Morphological, growing features, climatic conditions and storage conditions for picking and drying garlic may also influence the quantitative difference in the chemical composition of the garlic bulb (Lutomski, 1987). Some active ingredient, such as 5-allycysteins, are not present above trace levels in fresh garlic but are produced by aging chopped garlic in dilute alcohol. Other organosulfur compounds are produced through maceration in the edible oils. Another source of variability is the stability of garlic compounds in the different stages of storage between harvest and consumption (Singh and Singh, 1996). Common garlic preparations are as follows:

1.5.1 Fresh Garlic and Garlic Condiments

Home-prepared garlic and garlic condiments are common garlic preparation methods. The amount of chemical constituents in fresh garlic from different parts of the world varies and depends upon growing conditions. Garlic is also used in a variety of commercial conditions, such as garlic powder and garlic salt.

1.5.2 Commercial Preparations

A variety of commercial preparations of garlic is available today. A constituent in each formulation varies according to their way of preparation, as well as the bioavailability of the product after ingestion. Variation in constituents largely depends on manufacturing processes, origin of the garlic used and standardization methods.

1.5.3 Aged Garlic Extract

The manufacturer of Kyolic prepares garlic products using "Aged Garlic Extract™". It uses garlic bulbs that are aged up to 20 months, and it is used to prepare tablets, capsules and liquids. Consistency of preparation is maintained by measuring the amount of 5-allycysteins (SAC).

1.5.4 Dehydrated Preparations

Garlic powder is one of the most widely used commercial products. The dehydration of garlic is performed at a low temperature. The dried garlic is then pulverized and prepared into tablets. This garlic preparation is odor-free until it reaches the gastrointestinal tract after ingestion.

1.5.5 Distilled Oil of Garlic

In this method, crushed garlic is boiled in water, and steam is collected as it vaporizes.

1.5.6 Oil Macerates

Chopped garlic is mixed with edible oils. It is packaged in soft gel capsules.

1.5.7 Other Commercial Preparations

Various other garlic preparations are combination or enriched products with other herbs, minerals and vitamins. Some products include *Ginkgo biloba*, selenium, vitamin C and E, β-carotenes, calcium and enzymes.

REFERENCES

Abdullah TH, Kindil O, Elkadi A, Carter J (1988) Garlic revisited therapeutic for the major disease of our times? J Natl Med Assoc. 80(4): 439–445.

Alam K, Hoq O, Uddin S (2016) Medicinal plant *Allium sativum*: A review. *J Med Plants Stud.* 4(6): 72–79.

Batiha G, El-Saber Beshbishy AM, Wasef LG, Elewa YHA, AI-Sagar AA, Abd El-Hack ME, Toha AE, Abd-Ethakim YM, Devekota HP (2020) Chemical constituents and pharmacological activities of garlic (*Allium sativum*): A review. *Nutrients.* 12(872): 1–21. https://doi.org/10.3390nu12030872.

Bazaraliyeva A, Meldashov D, Turgambayeva A, Kartbayeva E, Kalykova A, Sarsenova L, Issayeva R (2022) Chemical and biological properties of bio-active compounds from garlic (*Allium sativum*). *Pharmacia.* 69(4): 955–964.

Block E (1985) The chemistry of garlic and onions. *Sci Am.* 252(3): 114–119.

Block E (1986) *In Folk Medicine: The Art and the Science.* Steiner, R.P., Ed. American Chemical Society, Washington, DC, pp. 125–137.

Block E (1992) The organosulfur chemistry of the genus Allium: Implications for the organic chemistry of sulfur. *Angew Chem Int Ed Engl.* 31: 1135–1178.

Block E (2010) *Garlic and other Alliums. The Lore and the Science.* The Royal Society of Chemistry. ISBN: 978-1-8473-180-5.

Bolton S, Null G, Troetel WM (1983) The magical uses of Garlic—fact and fiction. *Am Pharm.* 22: 40–43.

Christopher Ranch (1999) *Garlic: Good Taste and Good Health.* Available from URL: http//www.christopher-ranch.com/website5.htm[Web Page].

Day A (1999) *Gothic Gardening: Ye Oldegothick Herbal.* Hecate's Supper, Garlic. Available from URL: http:/www.gothicnet/~malice/garlic.html[Web Page].

Ernst E (1987) Cardiovascular effects of garlic (*Allium sativum*). A review. *Pharmatherapeutica.* 5: 83–89.

Farbman KS, Barnett ED, Bolduc GR, Klein JO (1993) Antibacterial activities of garlic and onions: A historical perspective. *Pediatr Infect Dis J.* 12: 613–614.

Fenwick GR, Hanley AB (1985a) The genus *Allium*, Part 2. *Crit Rev Food Sci Nutr.* 22: 273–377. https://doi.org/10.1080/10408398-509527417.

Fenwick GR, Hanley AB (1985b) The genus *Allium*, Part 3. *Crit Rev Food Sci Nutr.* 23(1): 1–73. https://doi.org/10.1080/10408398-509527419.

Hanley AB, Fenwick GR (1985). Cultivated alliums. *J Plant Foods.* 6: 211–238. http://doi.org/10.1080/0142968×1985.11904312.

Harris JC, Cottrell SL, Plumner S, Lloyd D (2001) Antimicrobial properties of (*Allium sativum*) Garlic. *Appl Microbial Biotechnol.* 57: 282–286.

Ipek M, Ipek A, Almquist SG, Simon PW (2005) Demonstration of linkage and development of the first low-density genetic map of garlic based on AFLP markers. *Theor Appl Genet.* 110: 228–236.

Kamenetsky R, Shafir IL, Khassanov F, Kik C, Van Heusden AW, Vrielink-Van Ginkel M, Burger-Meijer K, Augez J, Arnault I, Rabinowitch HD (2005) Diversity in fertility potential and organo-sulfur compounds among garlics from central Asia. *Biodivers Conserv.* 14: 281–285.

Kendler BS (1987) Garlic (*Allium sativum*) and onion (*Allium cepa*): A review of their relationship to cardiovascular disease. *Prev Med.* 16: 670–685.

Kleijnen J, Knipschild P, Ter Riet G (1989) Garlics, onions and cardiovascular risk factors, a review of the evidence from human experiments with emphasis on commercially available preperations. *Br J Clin Pharmacol.* 28: 535–544.

Koch HP, Hahn G (1988) *Knoblauch: Grundlagen der therapeutischen Awendung Von Allium sativum L.* Urban and Schwarzenberg, Munichen, Baltimore.

Koch HP, Lawson LD, Eds. (1996) *The Science and Therapeutic Application of Allium sativum L. and Related Species.* Williams and Wilkins, Baltimore, MD.

Lancaster JE, Boland M (1990) Flavor biochemistry. In *Onions and Allied Crops*, Vol. 3. Rabino HD, with Brewster JL, Eds. CRC Press, Boca Raton, FL, pp. 33–72.

Lau BHS, Keeler WH, Adetumbi MA (1983) Antifungal effect of garlic. *Ann Meet Am Soc Microbiol USA*. 387.

Lau BHS, Tadi PP, Tosk JM (1990) *Allium sativum* (garlic) and cancer prevention. *Nutr Res.* 10: 937–948.

Lawson LD (1993) *Bioactive Organosulfur Compounds of Garlic and Garlic Products (Role in Reducing Blood Lipids) in Human Medicinal Agents from Plants.* Kinghorn AD, Balandrin MF, Eds., 203rd meeting of the American Chemical Society, Washington, DC.

Lutomski J (1987) *Components and Biological Properties of Some Allium Species.* Published on the Occasion of the 40th Anniversary of the Institute of Medicinal Plants, 61–707, Pozan, Ul Libeeta, 27.

Morbidoni L, Arterburn M, Young V, Mullins D, Mulrow C, Lawrence V (2001) Garlic: Its history and adverse effects. *J Herb Pharmacother.* 1(1): 63–83.

Petkov V (1966) Pharmacological and critical studies of garlic. *Dtsch Apoth Ztg.* 106: 1861–1867.

Raghavan B, Abrahm KO, Shankaranarayana ML (1983) Chemistry of garlic and garlic products. *J Sci Ind Res.* 42: 401–409.

Reuter HD, Sendal A (1995) *Allium sativum* and *Allium ursinum*, Part 2, Pharmacology and medicinal application. *Phytomedicine.* 2: 73–91.

Roser D (1991) *Garlic for health. Garlic Reseearch Bureau, Cambridge.* Martinbooks/ Simon & Schuster, New York.

Shemesh E, Scholten O, Rabinowich HD, Kamenetsky R (2008) Unlocking variability: Inherent variation and developmental traits of garlic plants originated from sexual reproduction. *Planta.* 227: 1013–1024.

Singh RN, Kumar P, Kumar N, Singh DK (2020) Garlic (*Allium sativum*) pharmaceutical uses for human health. *Int J Pharm Sci Res.* 11(9): 4214–4228.

Singh VK, Singh DK (1996) Molluscicidal activity of pre and post harvest *Allium sativum* L. (Garlic). *Phytother Res.* 12: 311–318.

Singh VK, Singh DK (2008) Pharmacological effects of garlic (*Allium sativum* L.). *ARBS Ann Rev Biomed Sci.* 10: 6–26.

Sivasami K (2017) Scientometric analysis of garlic research publications. *Glob J Res Anal.* 6: 194–197.

Sprecher E (1986) *Allium sativum* L—Wundermittel oder Arzneipflanze? *Pharm Ztg.* 131: 3161–3168.

Srivastava KC, Bordia A, Verma SK (1995) Garlic (*Allium sativum*) for disease prevention. *S Afr J Sci.* 91: 68–77.

Srivastava KC, Mustafa T (1993) Pharmacological effects of species: Eicosanoid modulationg activities and their significance in human health. *Biomed Rev.* 2: 15–29.

Sumiyoshi H, Wargovich MJ (1989) Garlic (*Allium sativum*): A review of its relationship to cancer. *Asia Pac J Pharmacol.* 4: 133–140.

Tadu CK, Dutta T, Ghorai M, Biswas P, Samanta D, Oleksak P, Jha NK, Radha MK, Prockow J, de la Lastra JMP, Dey A (2022) Traditional uses, phytochemistry, pharmacology and toxicology of garlic (*Allium sativum*), a storehouse of diverse phytochemicals, a review of research from the last decade focusing on health and nutritional implications. *Front Nutr.* 1–19. https://doi.org/10.3389/fnut2022.929554.

Tellez TME, Aguilar ESV, Rojas MTA, Guinco DOD, Quevedo R, Diaz O, Montes JMB (2020) Garlic (*Allium sativum*) and its beneficial properties for health: A review. *Agroind Sci.* 10(1): 103–115.

Tesfaye A (2021) Revealing the therapeutic uses of garlic (*Allium sativum*) and its potential for drug discovery. *Hindwi Sci World J.* 2021, Article ID 8817288: 7. http://doi.org/10.1155/2021/8817288.

Tesfaye A, Mengesha W (2015) Traditional uses, phytochemistry and pharmacological properties of garlic (*Allium sativum*) and its biological active compounds. *Int J Sci Res Sci Eng Technol.* 1(5): 142–148.

Turner MJ (1990) Garlic and circulatory disorders. *J Roy Soc Health.* 110(3): 90–93. http://doi.org/10.1177/146642409011000306.

Volk GM, Henk AD, Richards CM (2004) Genetic diversity among U.S. Garlic clones as detected using AFLP methods. *J Am Soc Hortic Sci.* 129: 559–569.

Volk GM, Stern D (2009) Phenotypic characterstics of ten garlic cultivars grown at diverse North American locations. *Hortscience.* 44: 1238–1247.

Whitaker JR (1976) Development of flavor, odor and pungency in onion and garlic. *Adv Food Res.* 22: 73–133.

2 Chemistry of Garlic

2.1 HISTORY OF GARLIC CHEMISTRY

French Chemist M. Auguste Cahours, for the first time in the 1856 edition of *Leçons de Chimie Générale Élémentaire*, gave a brief description of the isolation of garlic oil using an alembic, a device developed by the Abrab alchemist Jabir ibn Hayyan in 800 CE. According to Cahours, bulbs of garlic in water produced a volatile oil with a pungent odor. This oil contained carbon, hydrogen and sulfur, probably monosulfide. Though it was not referenced in the book, the same work on the preparation and characterization of garlic oil was published by the German scientist Theodor Wertheim. He described it as allylschwefel (Block, 2010). Cahours with colleague Hofmann synthesized garlic oil from "alcool acrylique", a recently discovered compound. Cahours and Hofmann's work was apparent in their 1857 paper in the *Philosophical Transactions of the Royal Society* of London. Cahour and Hofmann described the published work (in 1844) of the German scientist Theodor Werthelim, who described garlic oil preparation as allylschwefel. According to Wertheim, his garlic oil was distilled at 140°C and contained 63.33% carbon and 8.80% hydrogen. The name allyl sulfur become diallyl sulfide, so the common chemical name allyl was established (Wertheim, 1844). Although earlier studies by Wertheim, Cahours and Hofmann on the chemistry of *Allium sativum* indicate its initial study, even earlier, less specific efforts to determine the chemistry of the garlic clove were mentioned in *Flora Española*, a Spanish work in 1762, "el ajo [garlic] en la analysis chymica" (Quer, 1762). In a 1814 book (Thornton, 1814), it was mentioned that garlic roots, among all the parts of the plant, have a strong penetrating and diffusible smell and an acrimonious, almost caustic, taste. The root is full of limpid juice, which constitutes almost one-fourth of its weight. It also loses one and a half times its weight by drying, having no smell or taste. Boiling the garlic bulb in water causes the loss of its all virtues. On distillation, it provides a small quantity of yellowish essential oil. This oil possesses the sensible qualities of garlic to an eminent degree. The alcoholic extract of the garlic bulb is oily and tenacious and precipitates metallic solutions. But the active ingredient is a thick, ropy essential oil, in which alone resided the smell, the taste and all that distinguishes garlic. According to Eric Block (2010), these descriptions of the chemical properties of garlic and its active ingredient allicin from almost 200 years ago were accurate based on current knowledge. FW Semmler (1892) found that the yellow-colored distilled oil of garlic, formed to the extent of 0.09% from the distillation of garlic bulbs at pressure of 16 mm, contained 60% diallyl disulfide, a smaller amount of diallyl trisulfide and tetrasulfide, which can be collectively known as diallyl polysulfides, as well as allyl propyl disulfide. The diallyl sulfide reported by Wertheim was absent (Semmler, 1892). The diallyl sulfide reported by Wertheim was formed by the decomposition of diallyl polysulfides at higher distillation temperatures, whereas Semmler used vaccum distillation at lower

DOI: 10.1201/9781003646211-2

temperatures so that temperature-sensitive garlic oil components were not decomposed. After five decades, a major development in the *Allium sativum* chemistry was noted in the year 1944. Chester J. Cavallito isolated a chemical from a chopped garlic clove, which had the chemical formula $C_6H_{10}S_2O$ and was named allicin. Cavallito extracted the product from the chopped garlic clove with ethanol and evaporated the ethanol at a temperature below 50°C with the help of a vacuum pump. In this way, Cavallito extracted 5 liters of ethyl alcohol from 4 kilograms of the garlic clove. After evaporation, he isolated 6 grams of a colorless liquid with a pungent odor. This oil was slightly denser than water, having 2.5% solubility and a molecular weight of 162. This product was different from diallyl disulfide and higher polysulfides, as noted by Semmler. Cavallito confirmed the structure of allicin by synthesis. He added an oxygen atom to the diallyl disulfide (1) with the help of the oxidizing agent perbenzoic acid (C_6H_5-Co_3H) (Figure 2.1).

Allicin (2) is a thiosulfinate or mono-oxidized disulfide, -S(O)S-. Other disulfides, such as dimethyl disulfide (3) and di-n-propyl disulfide (5), also undergo easy oxidation to the corresponding thiosulfinates (4) and (6), which also play important roles in garlic chemistry. According to Cavallito, thiosulfinates (4) and (6) can be easily distilled under vaccum and are more stable than allicin (2). Small et al. (1947) explained that the instability of the allyl compound was associated with the double bond structure. In contrast to allicin (2), thiosulfinate (4) is completely soluble in water. Cavallito was granted a U.S. patent for his process of extracting garlic and synthesizing allicin (Cavallito and Bailey, 1944a, 1944b; Cavallito et al., 1944, 1945; Cavallito and Small, 1950; Cavallito, 1951; Small et al., 1947, 1949). Later, Cavalltio noted that both natural and synthetic allicin demonstrated antibacterial activity, like penicillin. Allicin is inactivated by the amino acid cystine, forming S-allylmercaptocysteine (7).

In another experiment, Cavallito showed that allicin is a secondary compound formed by enzymatic action on precursors in the intact bulb. Garlic cloves were

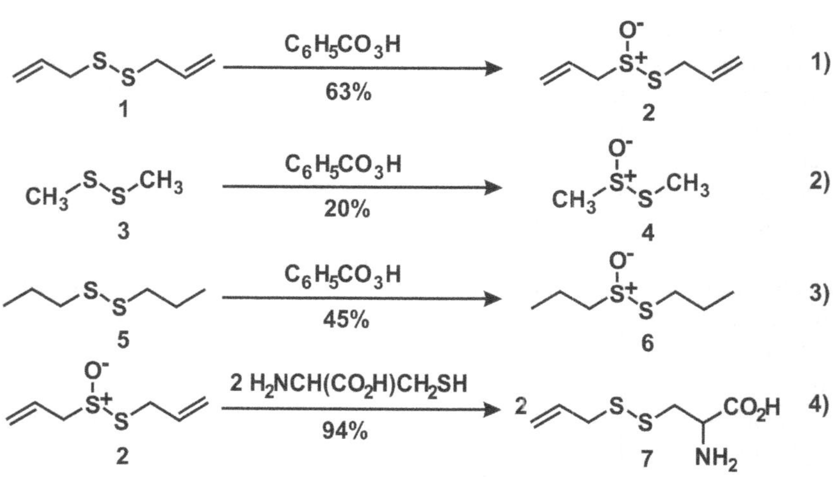

FIGURE 2.1 Thiosulfinates (2), (4) and (6) synthesis; reaction of allicin (2) with cysteine to form 7.

frozen in dry ice, pulverivzed and extracted with acetone. After the evaporation of acetone, only a small quantity of residue was yielded. No sulfides or allicin were noted. The white garlic powder was odorless, but with the addition of a small amount of water, the typical garlic odor was detected and isolated. This demonstrated that sulfides (allicin) are not found in the intact garlic bulb as much as they are found in garlic oil. Volatile substances of garlic are not found in the plant but results from disrupting the plant cells. The quantity of allicin is 70% of all thiosulfinates found in freshly crushed garlic, amounting to about 0.4% fresh weight. At room temperature, the half-life of pure allicin from a crushed garlic bulb is about 2.5 days. In water at 23°C and concentrations of 0.01 to 0.1%, the half-life of allicin is about one month, while at −70°C, it is indefinitely stable (Lawson, 1998). According to Fujisawa et al. (2008a, 2008b), a 0.1–0.2% solution of allicin in water has a half-life of about a year at 4°C, 32 days at 15°C and only one day at 37°C, and they indicate that ajoene is formed when allicin decomposes. Solvents such as water, which form hydrogen bonds with the allicin oxygen, retard the decomposition (Vaidya et al., 2009).

The precursor compound "alliin" is found in the intact garlic clove. When a garlic bulb is crushed, a vacuolar enzyme called alliinase is released, which breaks down alliin into very short-lived chemicals that form allicin through different chemical processes. Further reactions of allicin give rise to the diallyl polysulfides found in garlic oil (Figure 2.2).

Maximum yield of allicin from a garlic clove can be possible by special measures. The first step is that the garlic clove should be properly crushed/chopped by hand, then ground halfway to ensure maximum enzymatic yield and uniform homogenization of the pieces of garlic. Before grinding the tissue, the grinder should be chilled with an ice bath to avoid overheating the alliinase enzyme.

2.2 PUNGENCY OF CRUSHED GARLIC

Crushed/sliced garlic placed on the tongue and lips causes a painful burning sensation. Cut garlic can also cause irritation and inflammation of the skin, cornea and mucosa. This irritant behavior of cut garlic is due to the stimulation of nociceptors, which are neurons that respond to painful stimuli. These nociceptors, when in contact with the irritating chemicals of crushed garlic, send signals that can trigger inflammation, causing the perception of pain. Nociceptors are silent receptors and respond only when activated by irritating stimuli. These nociceptors use transient receptor potential (TRP) ion-channel proteins such as TRPA1 and TRPV-1. Activated TRP-1 mediates the flow of calcium ions into the ending of specialized neurons. Thus, ion flow excites the neurons, which causes local inflammation and pain. Chemicals found in crushed/sliced garlic cloves covalently attach themselves

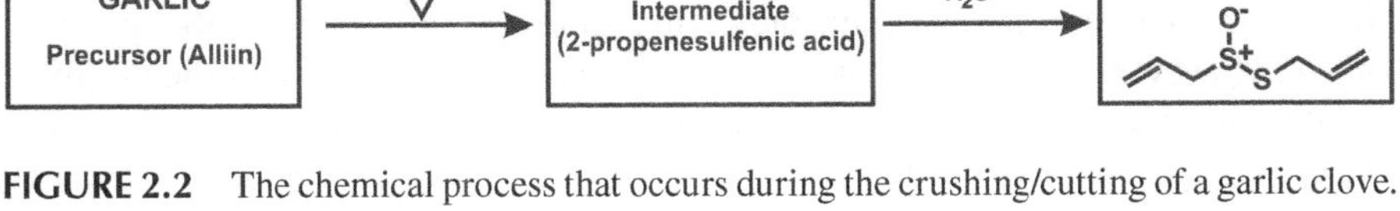

FIGURE 2.2 The chemical process that occurs during the crushing/cutting of a garlic clove.

to the thiol (–SH) group of cysteine amino acid residues on the channel protein (Caterina, 2007; Brone et al., 2008; Salazar et al., 2008). A second member of the TRP family, TRPV-1, also reacts with allicin and fresh garlic (Salazar et al., 2008). TRP-1 is ten times more sensitive to garlic and allicin than TRPV-1. Garlic extracts also mediate vasodilation through the activation of TRP-1. TRP-1 is well conserved across the animal kingdom, which suggests an ancestral role for this channel, probably in sensation (Garcia-Anoveros and Nagata, 2007). Many species including, ticks, mosquitoes and nematodes, are repelled by garlic, indicating the defensive role of its metabolites (Macpherson et al., 2005).

2.3 ODOR OF CUT GARLIC

The aroma of garlic is powerful, lingering and detectable, even on dilution. According to Laska et al. (2007), thiols and amines are major products of the microbial degradation of proteins and are involved in the putrefaction processes, which initiate the production of toxins. Primates are highly sensitive to such compounds in order to avoid intoxication. Day (1978) and Wang et al. (2003) suggested that metal ions are involved in olfactory receptors in the form of metalloproteins, which have a high affinity for sulfur and nitrogen compounds in garlic.

2.4 METABOLISM OF GARLIC COMPOUNDS

Pechey (1694), a British herbal scientist, noted that the application of garlic to the soles of the feet caused a garlic stink in the breath. According to him, garlic components mixed with the blood and were carried to the lungs, resulting in the garlic stink during expiration. Woodville (1793) mentioned in his book *Medical Botany* that each part of the garlic plant, more especially the bulb, has a pungent taste and strong smell. The bulb of garlic taken into the stomach is found in various excretions, such as urine, perspiration and even milk (Woodville, 1793).

Raw garlic compounds, when ingested in the body, are subjected to metabolic degradation. Germain et al. (2002, 2003) noted that rats orally fed with a single dose of 200 mg/kg dose of diallyl disulfide showed the presence of 2-propenethiol, allyl menthyl sulfide, sulfoxide and sulfone after 15 days. Teyssier et al. (1999) reported that diallyl disulfide is converted by the human liver microsomal enzyme P-450 to allicin.

2.5 GARLIC BREATH

Suarez et al. (1999) and Hasler (1999) reported that the consumption of garlic caused transient high concentrations of methanethiol and allyl mercaptan (2-propenthiol) and less concentrations of allyl methyl sulfide (AMS), allyl methyl disulfide and diallyl disulfide. Except AMS, all gases were present in greater concentrations in the mouth than in pulmonary air, indicating an oral origin; only AMS was of gut origin. However, after three hours of garlic consumption, AMS was the dominant breath sulfur gas. This higher concentration of AMS after three hours is due to the lack of gut and liver metabolism, whereas other gases are metabolized rapidly

(Taucher et al., 1996). This is the reason why the breath odor after ingesting garlic initially originates from the mouth and subsequently from the gut. Other studies done by various workers also support the earlier finding of organosulfur compounds found in human breath after the consumption of garlic (Cai et al., 1995; Taucher et al., 1996; Rosen et al., 2001; Lawson and Wang, 2005; Chen et al., 2007; Tamaki et al., 2008). Allicin and diallyl polysulfides are converted by plasma glutathione to 2-propenethiol, which is converted by the methyl transferanse enzyme with S-adenosylmethionine (SAM) to AMS. Organoselenium compounds were also found in the breath after the ingestion of raw garlic (Cai et al., 1995). Allicin formed in crushed garlic can easily pass through the skin, which explains the origin of garlic breath when fresh garlic is applied under the soles of the feet. If garlic is eaten with raw foods having deodorizing activities, thiols and disulfides are removed from the mouth; simultaneously, allyl methyl sulfide in the gut is also reduced. Brocklehurst (1953) claimed in an experimental and chemical study that chlorophyll can have a neutralizing effect on the odor of garlic.

MacDonald and Jacob (1928) noted that the ingestion of wild garlic by a cow or even the inhalation of garlicky fumes from the crushed plant in the field can be transmitted to the cow's milk. Granroth (1970) suggested vaccum treatment to remove garlicky odors and taste from the cow's milk. Although there are adverse effects on milk odor, there are also several benefits of garlic intake to cows and other ruminants. The intake of garlic reduced the methane production, a major source of greenhouse gases, without affecting rumen fermentation and digestion (Karma et al., 2006; Patra et al., 2006).

2.6 ALLICIN—PRECURSOR IN GARLIC BULB

Garlic clove chemistry began as earlier stated in this chapter. The next significant work on garlic chemistry was done by Arthur Stoll, who was a professor at the University of Munich and had completed his doctoral work on chlorophyll chemistry with 1915 Nobel Prize winner Richard Willstatter. He, along with his coworker Seebeck, turned his attention to garlic chemistry from 1947 to 1949. Both had identified the precursor reported in Cavallito's white powder as a nonprotein aminoacid, S-allyl-L-cysteine sulfoxide ($CH_2{=}CHCH_2S(O)CH_2CH\ (NH_2)CO_2H$). They called this compound alliin. Alliin undergoes enzymatic cleavage to allicin. During the extraction process, cleavage of the garlic clove was prevented. Garlic bulbs were deep frozen and cut into small pieces in frozen condition. Thereafter, it was extracted with methanol or ethanol and ethyl ether at low temperature. After fraction precipitation with ethyl alcohol, alliin is obtained in a pure state. According to Stoll and Seebeck (1948), 1 kg of fresh bulbs of garlic and 810 mg of alliin could be obtained, melting at 163.5°C. However, Lawson (1996) extracted 5 to 14 mg/g fresh weight alliin with the use of improved techniques. Seebeck was granted a U.S. patent for alliin (Seebeck, 1953).

After establishing the structure of alliin, Stoll and Seebeck worked on the synthesis of alliin from simple precursors, with the chiral natural amino acid (-)-L-cystenine, which react with allyl chloride, giving (-)-5-allyl-L-cysteine (deoxyallin) a natural compound. Oxidation of this compound yields alliin. Stoll and Seebeck later isolated and purified the garlic alliinase enzyme responsible for the conversion of alliin to allicin. According to Stoll and Seebeck (1951), alliinase reacts only with the derivatives of the natural amino acid L-cysteine. The alliinase enzyme has a stereo chemically well-defined cavity, which accepts natural alliin.

In garlic cells, alliin and alliinase enzymes are separated. The enzyme alliinase resides in the cell vesicle separated from alliin by their membranes. Crushing or injuring a garlic clove breaks the vesicular membrane, and alliinase comes in contact with alliin. Instantaneously, allicin is formed from the condensation of two molecules of 2-propenesulfenic acid. This reaction is very fast (> 97% complete after 30 s), as the cellular level of alliin and alliinase are equal. Enzyme alliinase (EC 4.4.1.4) belongs to the glycoprotein family, which contain 6% carbohydrate. Kuettner et al. (2002) and Shimon et al. (2007), in the X-ray crystallography study of alliinase, demonstrate that it is a dimer composed of two equal subunits, each containing 448 amino acids, having a total molecular mass of 10,3000 daltons.

Alliinases in garlic cloves are localized in a vascular bundle sheath inside the cells, which surrounds the veins, while alliin and other related cysteine derivatives are concentrated in the mesophyll cells (Ellmore and Feldberg, 1994; Yamazaki et al., 2002). In this way, the presence of alliinase in the vascular bundle sheath is helpful in protecting against microorganism or small herbivore invasion through the veins into mesophyll cells. Alliinase is activated by a cofactor, pyridoxal-5-phosphate or "PLP". When alliinase is not in use, it exists as a complex known as an internal aldimine, involving the 251st amino acid (lysine 251). Alliin, having a specific three-dimensional shape, fit precisely into an enzymatic cavity. On contact, alliin substitutes for lysine 251, forming an external aldimine. When alliin is activated as an aldimine, it loses a proton to a basic site in the enzyme. In the next step, the alliin carbon-sulfur bond on the amino acid side undergoes cleavage (an α, β elimination reaction), releasing 2-propenesulfenic acid and leaving behind a PLP-bound amino-acryl intermediate. Alliinase is called a C-S lyase enzyme. The intermediate product of cleavage, 2-propenesulfenic acid (CH_2=$CHCH_2S$-O-H), has never been isolated, even at very low temperatures. It rapidly self-condenses with the loss of water to form allicin (Figure 2.3) (Block, 2010).

The PLP-bond aminoacrylate undergoes rapid hydrolysis, giving pyruate and ammonia as lysine 251 is reinstalled. The system functions like a very efficient assembly line, in which alliin enters in one end and allicin, ammonia and pyruvate come out at the other end, while the alliinase remains unchanged. The overall process is so rapid that a burst of allicin is produced upon mixing precursor and enzyme (Block, 2010). Allicin is unstable and reactive. It can be only stored unchanged for extended periods as a refrigerated aqueous solution.

2.7 THIOSULFINATES IN GARLIC

Garlic contains at least 100 sulfur-containing compounds fundamental to its medicinal uses (Lawson, 1993; Cai et al., 1994; Srivastava et al., 1995). Allicin represents 70–80% of the total thiosulfinates formed (Figure 2.4)

Other thiosulfinates include allyl-ss(o)-methyl (6–16% of total), methyl-ss-o-allyl (3–9% of total), trans-1-propenyl-ss(o)-methyl+methyl-ss(o)-trans-1propenyl (0.1–2.5%

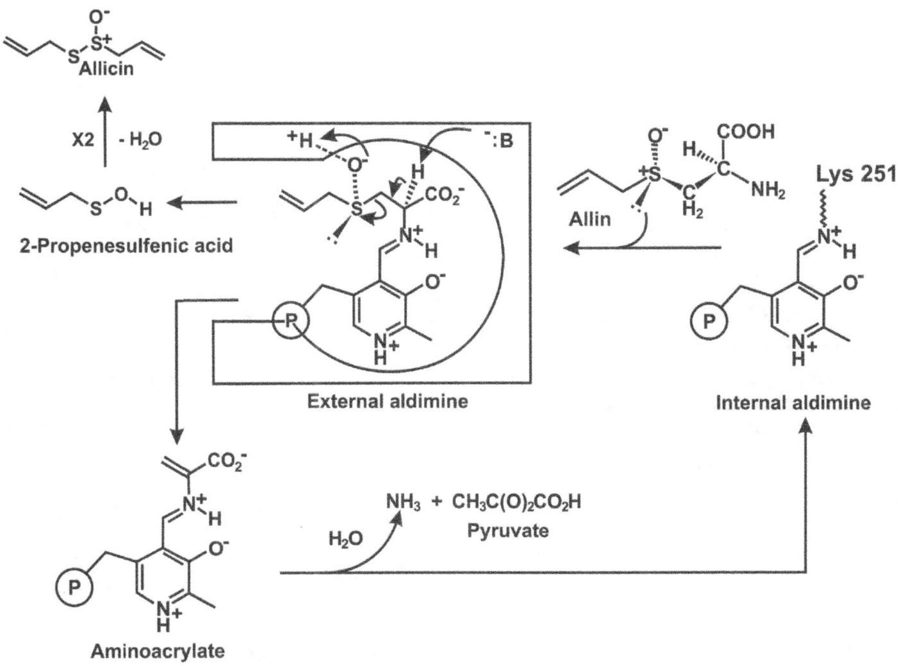

FIGURE 2.3 Activity of alliinase on alliin: In the absence of substrate, alliinase forms a covalent attachement with Lys251, the internal aldimaine. In presence of the substrate alliin, alliinase binds covalently to the PLP (pyridoxal-5'-phosphate) to form an external aldimine. The carbon-sulfur bond is broken, 2-propenesulfienic acid is formed, which joins to form allicin, leaving aminoacrylate bound to the PLP. This aminoacrylat is lysed from the PLP and hydrolyzed to form pyruvic acid and ammonia. Now the PLP is free to reform the internal aldimine with Lys251.

of total) and methyl-ss(o)-methyl (2% of total) thiosulfinates (Lawson et al., 1991a, 1991b). The formation of allicin (allyl-2-propene thiosulfinate or diallyl thiosulfinate) is completed in 0.2 to 0.5 min at room temperature. The formation of the methyl thiosulfinate takes 1.5 to 5.0 min (Lawson and Hughes, 1992). All possible combinations of 2-propene, 1-propene and methane sulfenic acids result in thiosulfinates (Lawson et al., 1991a, 1991b).

The sulfinates released from crushed garlic are reactive molecules and undergo a number of transformations, depending on the temperature, pH and solvent conditions. Allyl-S-thiosulfinates (allicin) are the least stable of eight thiosulfinates (Block, 1992; Lawson and Block, 1997). The half-life of allicin (concentration of 0.1–0.4 mg/ml) at room temperature is ten days in 1 mM citric acid (pH3), four days in water, 48 hours in methanol or chloroform, 24 hours in hexane and three hours in ether (Lawson, 1993).

The principal transformation products after the incubation of thiosulfinates in water are diallyl trisulfide, diallyl disulfide and allyl methyl trisulfide (Block, 1992). The incubation of allicin or allyl methane thiosulfinate in low polarity solvents (or without solvent) primarily produces 1, 3-vinyldithiin (2-vinyl-4H-, 3dithiin) (51% of

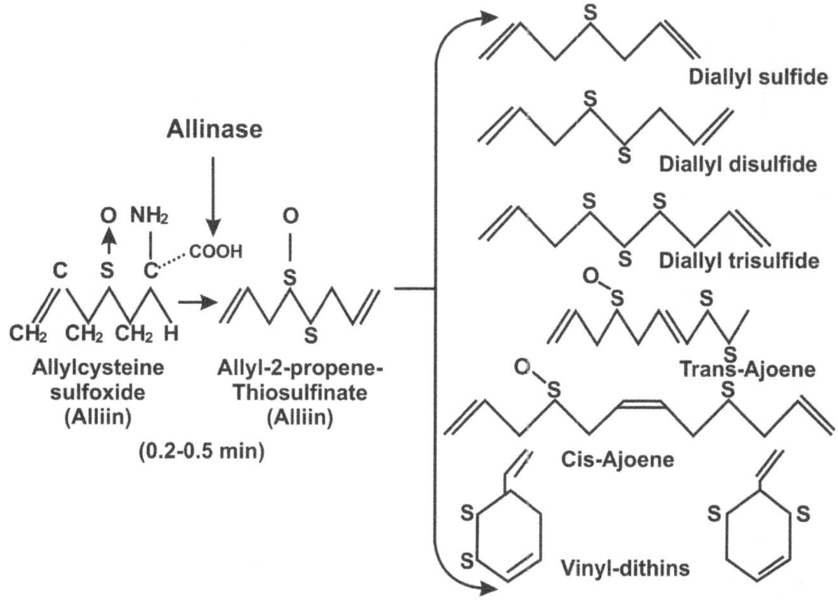

FIGURE 2.4 Synthesis of allicin from alliin by the enzyme alliinase.

total), 1–2 V-venyldithiin (3-vinyl-4H-1, 2-dithiin) (19% of total) and lesser amounts of ajoene (E, Z-4, 5, 9-trithiododeca-1, 6, 11-trien 9 oxide) (12% of total) and sulfides (18% of total). The incubation of pure allicin in methanol yields 8% diallyl trisulfide and about equal amounts (20–25% of total) of ajoene and the vinyldithiins (Lawson et al., 1991a, 1991b; Lawson, 1993). Steam distillation of garlic bulbs yields garlic oil that consist of diallyl (57% of oil), allyl methyl (37% of oil) and dimethyl (mono to hexa sulfides) (6% of oil of total) (Lawson, 1993). The essential oil content of fresh garlic bulb is between 0.09% and 0.35% (Lutomski, 1987).

2.8 EFFECT OF COOKING ON GARLIC COMPOUNDS

The effect of cooking on the organosulfur compounds in garlic bulb is an important concern and common question. Boiling of garlic cloves for 20 min causes a 0.12% loss of the ϒ-glutamyl cysteines to S-allyl cysteine and S-1-propenyl cysteine and a similar loss of allicin without the production of S-allycysteine (Lawson, 1993). When garlic is chopped into tiny pieces or is crushed, most of the cysteine sulfoxides are transformed into thiosulfinates. Boiling crushed garlic for 20 min in a closed container causes the complete conversion of the thiosulfinates to diallyl trisulfide and lesser amounts of diallyl disulfide and allyl methyl trisulfide. Cooking in an open container for 10 to 20 min retained 21% to 7% of the thiosulfinate, respectively, but 97% of the sulfides had evaporated. Cooking in milk in an open container causes an even more rapid loss of allicin and the other thiosulfinates (only 0.5% remained after 10 min), but 70% of the sulfides formed were retained, even after 40 min, probably by the milk fat (Lawson, 1993).

2.9 GARLIC PRODUCTS

Different products of garlic are used in preharvest and postharvest conditions. In preharvest conditions, the chemical composition of garlic depends on genotype, climate requirement, growing condition, irrigation, fertilization and harvesting stage. Postharvest conditions point out processing methods, consumption forms (aged garlic, fresh garlic, cooked garlic, dried garlic, etc.), humidity, harvesting time and storage time of the garlic clove after harvesting (the relationship is shown in Figure 2.5) (Singh and Singh, 1996; Martins et al., 2016).

2.9.1 Garlic Powder

Garlic powder is simply dehydrated, pulverized garlic bulbs; their chemical composition and alliinase activity is the same as fresh garlic, if processed carefully. Most garlic powders are prepared by drying chopped garlic at 50°C–60°C. Some garlic powders are prepared by spray-drying homogenized garlic, resulting in a complete loss of alliin and allicin. Garlic powder has also been prepared by freeze-drying garlic to avoid possible heat damage (Singh and Singh, 1995). Oven drying at 60°C

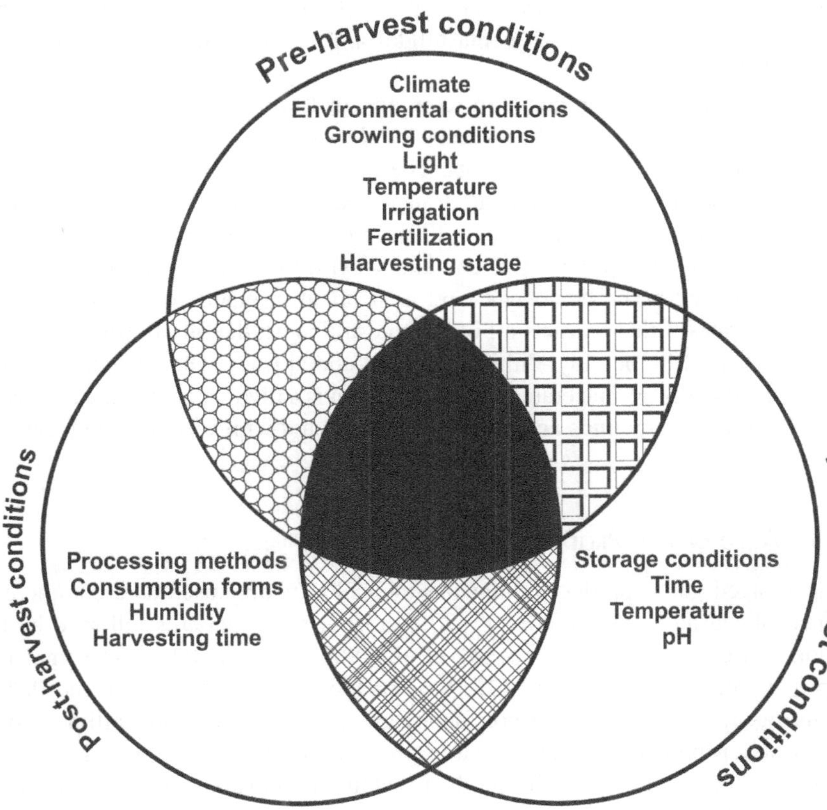

FIGURE 2.5 Chemical contents of the garlic clove in pre- and postharvest conditions.

does not affect the yield of the main thiosulfinates (allicin and the allyl methane thiosulfinates (Lawson and Hughes, 1992; Lawson, 1993)). Garlic powder products are enteric coated to prevent the rapid inhibition of alliinase by stomach acid (Lawson and Hughes, 1992).

2.9.2 Garlic Oil

There are three types of garlic oil:

1. The oil of the steam-distilled crushed garlic.
2. The oil of crushed garlic macerated in a vegetable oil.
3. The oil produced by ether extraction of the thiosulfinates of crushed garlic.

All these oils are not present in garlic cloves. Garlic contains a small amount of (0.2%) oil in garlic cloves—mainly triglycerides, phospholipids and glycolipids (Fenwick and Hanley, 1985).

Steam-distilled garlic oil consists of the diallyl (57%), allyl methyl (37%) and dimethyl (6%) mono to hexasulfides. Commercial steam-distilled garlic oil contains 26% diallyl disulfide, 19% diallyl trisulfide, 15% allyl methyl trisulfide, 13% allyl methyl disulfide, 8% diallyl tetrasulfide, 6% allyl methyl tetrasulfide, 3% dimethyl trisulfide, 3% monosulfides, 4% pentasulfides and 1% hexasulfides (Lawson et al., 1991a, 1991b).

Oil-macerated garlic products are common in Europe. They are produced by homogenizing chopped garlic in a vegetable oil (soyabean oil), followed by separation from the oil-insoluble components. The thiosulfinates generated during chopping are very soluble in the oil and are transformed to vinyl-dithiins (70%), sulfides (18%) and ajoenes (12%) (Lawson et al., 1991a, 1991b). Oil-macerated garlic oil is the only commercial garlic product that contains vinyl-dithiins and ajoenes.

Ether-extracted garlic homogenate oil is identical in qualitative composition to the oil macerates. Ether extracts all of the thiosulfinates; the oily thiosulfinates are rapidly converted to the vinyl-dithiins, sulfides and ajoenes. The advantage of this oil, compared to the oil-macerate products, is that it is a highly concentrated preparation of the same compounds and can be further diluted with vegetable oil (Lawson, 1993).

2.9.3 Aged Garlic Product

In Japan, aged garlic product in aqueous alcohol is widely used as a commercial product. Sliced garlic is stored in 15–20% ethanol for 20 months followed by filteration and concentration (Nakagawa et al., 1980; Lawson, 1993). The final liquid product contains 10% W/V, 12% V/V ethanol, 0.05% total sulfur and 0.29% total nitrogen. This product does not contain cysteine sulfoxide or any alliin-derived transformation product, as these volatile products are lost in the long incubation period (Lawson et al., 1991b). This product contains a small amount of ϒ-glutamyl-S-allylcysteine (0.28 mg/g), ϒ-glutamyl-S-1propenylcysteine (0.17 mg/g), ϒ-glutamyl-S-methylcysteine (0.02 mg/g) and their hydrolysis products, S-allylcysteine (0.30

mg/g), S-1-propenylcysteine (0.15 mg/g), S-methylcysteine (0.11 mg/g), but no cysteine or glutathione.

2.10 SELENIUM IN GARLIC CLOVES

Selenium, an element of the same group as sulfur, plays an important role in biology. It has long been known as an essential micronutrient, whose dietary absence causes skeletal and cardiac muscle dysfunction. Selenium is required for the proper functioning of the immune system and prevention of cancer and premature aging. The selenium content of the garlic cloves is extremely low, 0.01–0.2 μg/g fresh weight. Recommended dietary allowance of selenium is 70 μg/day (equivalent to 1.0 kg garlic/day) (Lawson, 1993). The selenium content of garlic has been increased to 50 μg/g by cultivation in selenium-enriched soil for the purpose of improving the antitumor effect of garlic (Ip et al., 1992). Known selenium-dependent enzymes contain selenocysteine, a cysteine analog in which selenium replaces sulfur. Seleocysteine, called the 21st amino acid and essential for ribosome-directed protein synthesis, is incorporated into protein by tRNA directed by a UGA Codon (Axley et al., 1991; Hatfield et al., 2006). More than 25 selenoproteins are known (Papp et al., 2007).

REFERENCES

Axley MJ, Bock A, Stadtman TC (1991) Catalytic properties of an *Escherichia coli* formate dehydrogenase mutant in which sulfur replaces selenium. *Proc Natl Acad Sci USA*. 88: 8450–8454.

Block E (1992) The organosulfur chemistry of the genus Allium—implication for the organic chemistry of sulfur. *Angew Chem Int Ed Engl*. 31: 1135–1178.

Block E (2010) *Garlic and other Alliums, The Lore and the Science*. Published by The Royal Society of Chemistry, Thomas Graham House, Science Park, Milton Road, Cambridge, ISBN: 978-1-84973-1805.

Brocklehurst JC (1953) An assessment of chlorophyll as deodorant. *Br Med J*. 4809: 541–544.

Brone PJP, Morrannes R, Mercken M, Nuydens R, Meert T, Gijsen HJ (2008) Tear gasses CN, CR and CS are potent activators of the human TRPA1 receptor. *Toxicol Appl Pharmacol*. 231: 150–156.

Cahours MA (1856) *Lecons de Chimie Generale Lementaire*. Mallet-Bachelier, Paris.

Cahours MA, Hofman AW (1856–1857) Note on a new class of alcohols. *Proc R Soc, London*. 8: 33–34.

Cai X-J, Block E, Uden PC, Quimby BD, Sulliman JJ (1995) *Allium* Chemistry: Identification of natural abundance organoselenium compounds in human breath after ingestion of garlic using gas chromatography with atomic emission detection. *J Agric Food Chem*. 43: 1751–1753.

Caterina MJ (2007) Chemical biology sticky spices. *Nature*. 445: 491–492.

Cavallito CJ (1951) *Extraction of Garlic*. U.S. Pat US 2554088.

Cavallito CJ, Bailey JH (1944a) Allicin, the antibacterial principle of *Allium sativum* isolation, physical properties and antibacterial action. *J Am Chem Soc*. 66: 1950–1951.

Cavallito CJ, Bailey JH (1944b) Preliminary note on the inactivation of antibiotics. *Science*. 66: 390.

Cavallito CJ, Bailey JH, Buck JS (1945) The antibacterial principle of *Allium sativum*. III its precursor and essential oil of garlic. *J Am Chem Soc*. 67: 1032–1033.

Cavallito CJ, Buck JS, Suter CM (1944) Allicin, the antibacterial principle of *Allium sativum*-II Determination of the chemical structure. *J Am Chem Soc*. 66: 1952–1954.

Cavallito CJ, Small LVD (1950) *Hydrocarbon Esters of Hydrocarbonylthiol Sulfinic Acids and Their Process of Preperation*. U.S. Pat US 2508745.

Chen H, Wortmann A, Zhang W, Zenobi R (2007) Rapid *in vivo* fingerprinting of nonvolatile compounds in breath by extractive electrospray ionization quadrupole time of flight mass spectrometry. *Angew Chem Int Ed*. 46: 580–583.

Day JC (1978) New nitrogen bases with severe steric hindrance due to flanking tert-butyl groups, CIS-2, 6-di-tert-butylpiperidine. Possible steric blocking of olfaction. *J Org Chem*. 43: 3646–3649.

Ellmore GS, Feldberg RS (1994) Alliin lyase localization in bundle sheath of the garlic clove (*Allium sativum*). *Ann J Bot*. 81: 89–94.

Fenwick GR, Hanley AB (1985) The genus Allium—Part I. *Crit Rev Food Sci*. 22: 199–271.

Fujisawa H, Suma K, Origuchi K, Kumagai H, Seki T, Ariga T (2008a) Biological and chemical and stability of garlic derived allicin. *J Agric Food Chem*. 56: 4229–4235.

Fujisawa KS, Origuchi K, Kumagai H, Seki T, Ariga T (2008b) Thermostability of allicin determined by chemical and biological assays. *Biosci Biotechnol Biochem*. 72: 2877–2883.

Garcia-Anoveros J, Nagata K (2007) TRPA1. *Handb Exp Pharmacol*. 179: 347–362.

Germain E, Auger J, Ginies C, Siess MH, Teyssier C (2002) *In vivo* metabolism of diallyl disulphide in the rat: Identification of two new metabolites. *Xenobiotica*. 32: 1127–1138.

Germain E, Chevalier J, Siejs MA, Teyssier C (2003) Hepatic metabolism of diallyl disulphide in rat and man. *Xenobiotica*. 33: 1185–1199.

Granroth B (1970) Biosynthesis and decomposition of cysteine derivatives in onion and other *Allium* species. *Ann Acad Sci Fenn Chem*. 154: 1–71.

Hasler WL (1999) Garlic breath explained: Why brushing your teeth won't help? *Gastroenterology*. 117: 1248–1250.

Hatfield DL, Carlson BA, Xu X-M, Mix H, Gladyyshev VN (2006) Selenocysteine incorporation machinery and the role of selenoproteins in development and health. *Prog Nucleic Acid Res Mol Biol*. 81: 97–142.

Ip C, Lisk DJ, Stoewsand GS (1992) Mammary cancer prevention by regular garlic and selenium-enriched garlic. *Nutr Cancer*. 17(3): 279–286. http.//doi.org/10.1080/01635589209514197.

Karma DN, Agarwal N, Chaudhary LC (2006) Inhibition of rumen methanogensis by tropical plants containing secondary metabolites. *Int Cong Ser*. 1293: 156–163.

Kuettner EB, Hilgenfield R, Weiss MS (2002) The active principle of garlic and atomic resolution. *J Biol Chem*. 277: 46402–46407.

Laska M, Bautista MR, Hofelmann D, Sterlemann V, Salazar LTH (2007) Olfactory sensitivity for putrefaction—associated thiols and indoles in three species of non-human primate. *J Exp Biol*. 210: 4169–4178.

Lawson LD (1993) *Bioactive Organosulfur Compound of Garlic and Garlic Products in Human Medicinal Agents from Plants*. Kinghorn AD, Balandrin MF, Eds., ACS Symposium Series 534. American Chemical Society, Washington, DC, Chapter 21, pp. 306–330.

Lawson LD (1996) Garlic. In *The Science and Therapeutic Applications of Allium sativum L, and Related Species*. Koch HP, Lawson LD, Eds. Williams and Wilkins, Baltimore, MD.

Lawson LD (1998) A review of its medicinal effects and indicated active compounds. In *Phytomedicines of Europe: Chemistry and Biological Activity*. Lawson LD, Bauer R, Eds., ACS Symposium Series 691. American Chemical Society, Washington, DC, pp. 176–209.

Lawson LD, Block E (1997) Comments on garlic chemistry stability of S-(2-propenyl) 2-propene-1-sulfinothioate (allicin) in blood solvents and stimulated physiological fluids. *J Agric Food Chem.* 45(2): 542.

Lawson LD, Hughes BG (1992) Characterization of the formation of allicin and other tholulfinates from garlic. *Planta Med.* 58: 345–350.

Lawson LD, Wang ZJ (2005) Allicin and allicin—derived garlic compounds increase breath acetone through allyl methyl sulfide: Use in meascring sllicin bio availability. *J Agric Food Chem.* 53: 1974–1983.

Lawson LD, Wang ZYJ, Hughes BG (1991b) Identification and HPLC quantitation of the sulfides and dialk(en)yl thiosulfinates in commercial garlic products. *Planta Med.* 57: 363–370.

Lawson LD, Wood SG, Hughes BG (1991a) HPLC analysis of allicin and other thiosulfinates in garlic clove homogenates. *Planta Med.* 57: 263–270.

Lutomski J (1987) Compounds and biological properties of some Allium species. Published on the occasion of the 40th anniversary of the institute of medicinal plants. *Poznan Ul Libelta.* 27: 61–707.

MacDonald MB, Jacob M (1928) Inhalation the chief factor in onion or garlic contamination of milk. *Science.* 68: 568–569.

Macpherson LJ, Geierstanger BH, Viswanath V, Bandell M, Eid SR, Hawang S, Patapoutian A (2005) The pungency of garlic: Activation of TRAA1 and TRPV1 in response to allicin. *Curr Biol.* 15: 929–934.

Martins N, Petropoulos S, Ferreira IC (2016) Chemical composition and bioactive compound of garlic (*Allium sativum*l) as affected by pre and post harvest condition: A review. *Food Chem.* 21: 41–50. http.//doi.org/10.1016/J.Foodchem.216.05.0.29.

Nakagawa S, Masamoto K, Sumiyoshi H, Kunihiro K, Fuwa T (1980) Effect of raw and extracted—aged garlic juice their organs after preoral administration. *J Toxicol Sci.* 5: 91.

Papp LV, Lu J, Holmgren A, Khanna KK (2007) From selenium to selenoproteins: Synthesis, identify and their role in human health. *Antioxid Redox Signal.* 9: 775–806.

Patra AK, Karma DN, Agarwal N (2006) Effect of species on rumen fermentation methanogensis and protozoa counts in *in vitro* gas production test. *Int Congr Ser.* 1293: 176–179.

Pechey J (1694) *The English Herbal of Physical Plants.* Reprinted by Medical Publications Ltd., London, 1951.

Quer DJ (1762) *Flora Espanola, o histroia de las plantas que se crian en Espana.* Tomo Segundo, Madrid, Joachin, Ibarra.

Rosen RT, Hiseroolt RD, Fukuda EK, Ruiz RJ, Zhou Z, Lech J, Rosen SL, Hartman TG (2001) Determination of allicin, S—allylcysteine and volatile metabolities of garlic in breath, plasma or simulated gastric fluids. *J Nutr.* 131: 9685–9715.

Salazar H, Llorente I, Jara-Oseguera A, Garcia-Villegas R, Munari M, Gordon SE, Islas LD, Rosenbaum T (2008) A single n-terminal cysteine in TRPV1 determines activation by punget compounds from onion and garlic. *Nat Neuro Sci.* 11: 255–261.

Seebeck E (1953) *Crystalline Derivative from Allium Plants.* US 2642374.

Semmler FW (1892) Essential oil of garlic (*Allium sativum*). *Arch Pharm.* 230: 434–443.

Shimon LJW, Rabinkov A, Shin I, Mison T, Morelman D, Wilchek M, Frolow F (2007) Two structures of allinase from *Allium sativum* L: Apo form and ternary complex with amino acrylate reaction intermediate covalently bound to the PLP cofactor. *J Mol Biol.* 366: 611–625.

Singh VK, Singh DK (1995) Characterization of allicin as molluscicidal agent in *Allium sativum* (Garlic). *Biol Agric Hortic.* 12: 119–131.

Singh VK, Singh DK (1996) Molluscicidal activity of pre and post harvest *Allium* sativum (Garlic). *Biol Agric Hort.* 12: 311–318.

Small LD, Bailey JH, Cavallito CJ (1947) Alkyl thiosulfinate. *S J Am Chem Coc.* 69: 1710–1713.

Small LD, Baily JH, Cavallito CJ (1949) Comparison of some properties of thiosulfinates and thiosulfonate. *J Am Chem Soc.* 71: 3565–3566.

Stoll A, Seebeck E (1948) Allium compounds, I. Alliin, the true mother compound of garlic oil. *Helv Chim Acta.* 31: 189–210.

Stoll A, Seebeck E (1951) Chemical investigations on allicin, the specific principle of garlic. *Adv Ezymol.* 11: 377–400.

Suarez F, Springfied JF, Levitt M (1999) Differentiation of mouth versus gut site of origin of odoriferous breath gases after garlic ingestion. *Am J Physiol.* 276: 187–199.

Tamaki K, Sonoki S, Tamaki T, Ehara K (2008) Measurement of odour after *in vitro* or *in vivo* ingestion of raw or heated garlic, using electronic nose, gas chromatography and sensory analysis. *Int J Food Sci Technol.* 43: 130–139.

Taucher J, Hansel A, Jordan A, Lindinger W (1996) Analysis of compounds in human breath after ingestion of garlic using proton transfer—reaction mass spectrometery. *J Agri Food Chem.* 44: 3778–3782.

Teyssier C, Guenot L, Suschetet M, Siess MH (1999) Metabolism of diallyl disulfide by human liver microsomal cyctochromes P-450 and flavin-containing mono oxygenases. *Drug Metab Dispos.* 27: 835–841.

Thornton RJ (1814) *A Family Herbal*, 2nd ed. R and R Crosby, London.

Vaidya V, Ingold KV, Pratt DA (2009) Garlic: Source of the ultimate antioxidants—sulfenic acids. *Angew Chem Int Ed.* 48: 157–160.

Wang J, Luthey-Schulten ZA, Suslick KS (2003) Is the olfactory receptor a metalloprotein? *Proc Natl Acad Sci USA.* 100: 3035–3039.

Wertheim T (1844) Investigations on garlic oil. *Ann Chem Pharm.* 51: 289–315.

Woodville W (1793) *Medical Botany.* James Phillips, London.

Yamazaki M, Sugiyama M, Soito K (2002) Intercellular socialization of cysteine synthase and alliinase in bundle sheaths of Allium plants. *Plant Biotechnol.* 19: 7–10.

3 Garlic Uses in Folklore

The scientific study of traditional plant medicines can be considered a major part of ethenopharmacology, a term that was introduced in 1967. This is traditional knowledge that are restricted to a group that has lived in a locality for a long period of time but does not have a sophisticated theoretical framework, formal education and documented written history. Some ethanographers would distinguish between folklore and ethanopharmacology, claiming that the former is common knowledge in the population as a whole, largely concerning remedies for minor conditions, based on relatively innocuous material. Ethanopharmacology is more concerned with the knowledge of a few specialists, who are regarded by society as able to correctly diagnose and treat diseases, usually using more potent products. In many situations, these specialists are also linked with the religious practices of the society.

3.1 HISTORICAL DIMENSION OF GARLIC IN FOLK MEDICINE

Among early writings on the uses of garlic, the *Codex Ebers* is an Egyptian holy book (1550 BCE) that explains more than 875 therapeutic formulas, of which 22 contain garlic. The use of garlic is also given in the "Greek magical papyri" (2nd century BCE to the 5th century BCE) (Betz, 1992). In addition, garlic is also used by the Egyptians in embalming and mummification (Nicholson and Shaw, 2000; Manniche, 2006).

The most comprehensive compilations of remedies associated with garlic are found in Pliny the Elder's (Gaius Plinius Secundus) the *Natural History* (77 CE), which has been translated many times (Rackham, 1971). One of the most spectacular editions was translated into Florentine Italian by Cristofora Landino and illuminated, bound and printed in Venice by Nicolous Jenson around 1480. These books' volumes are the most treasured volumes of Oxford University's Bodlein Library. Book 19 of the *Natural History* deals with cultivation and botanical characteristics of garlic, onion, leek, chives and several wild garlics; whereas, book 20 deals with remedies derived from these plants.

In volume 20, Chapter 23, Pliny describes the 61 different remedies. He writes:

> Garlic has powerful properties and is of great benefit utility against changes of water and residence. Garlic keeps off serpents and scorpions by its smell it cures bites when drunk or eaten or applied as ointment. It is an antidote against the poisonous bite of the shrew-mouse pounded garlic has been given in milk to asthmatics. The ancients used also to give it raw to mad men.

Dioscorides was a military surgeon under the Roman Emperor Nero. He wrote five volumes in Greek on the medicinal values of plants. His observations and studies were based on his extensive travel in Southern Europe and Northern Africa. Later, this book was translated into Latin as *De materia medica*. This book is one of the authoritative works on medicinal plants for 1,500 years. *De materia medica* is one of the important

DOI: 10.1201/9781003646211-3

books on the history of herbal science; it also gives us knowledge of the herbs and remedies used by the Greeks, Romans and other cultures of antiquity. This work presents about 500 plants. There is an extensive discussion on garlic. Dioscorides describes 23 medicinal uses, including "Cleaning the arteries", driving out intestinal parasites, acting as a mild diuretic and as an antidote for various poisons and "injurious change in water" when taken internally (diarrhea, amoebic dysentery). Garlic can also be used externally for various skin afflictions. John Goodyer (1652 to 1655) translated this book in English (Gunther, 1933). Dioscorides writes about garlic:

> it hath a sharp, warming biting quality, expelling of flatulencies and disturbing of the belly, and drying of the stomach causing of thirst. . . . dulling of sight of the eyes and drawes away the urine. . . . Being dranck with decotion of origanum, it doth kill lice and nitts. . . . Kept in the mouth it does assuage the paine of ye theeth. . . . It open ye mouths of ye veins.
>
> **(Gunther, 1933)**

In first half of the 6th century, the Bower Manuscript was written, which also includes a group of medical texts (Wujastyk, 2003). In 1890, British Lieutenant Hamilton Bower obtained the manuscripts in a Buddhist monastery at the old Silk Road trading stop of Kunqa, China. This is the geographic point where the Kazakhstan, Kyrgyzstan and China border meet. After translating this manuscript, Bower sold it to the Bodlein Library with his name. Three parts of the manuscript deal with ayurvedic medicines (practiced in India for more than 2,000 years), written in Sanskrit. It is belived that the manuscript was copied by Buddhist monks. The Bower Manuscript includes the mythical origin of garlic and its medical uses in 43 verses.

In his writing, the Bower Manuscript describes how those who were forbidden to eat garlic could consume it indirectly by feeding it to a cow and then consuming the animal's milk products:

> When a cow has been kept waiting for three nights with almost no grass, a preparation of two part grass to one part garlic stalks is provided to cow. A Brahmin can then part take of her milk, curds, ghee or even butter milk and banish various diseases while maintaining propriety.
>
> **(Wujastyk, 2003)**

According to the Bower Manuscript, garlic is a "universal remedy" that does the following:

> drives away pallid skin disease, appetite-loss, abdominal lumps, coughs, thinness, leprosy, and weak digestion. It removes wind, irregular periods, gripes, phthisis, stomach-ache, enlarged spleen and piles. It takes away paralysis of one side, lumbago, worm disease, colic and urinary disorders. It completely conquers lassitude/catarrh, rheumatism of the arms or back and epilepsy. . . . I have declared the use of garlic as it was seen in older times by the sages. One should practice it properly.

A discription of the spring garlic festival is given in the manuscript. In this festival, garlands of garlic are displayed in front of houses and on the gates and are worn by the occupants of the house. This type of worship is performed in the courtyard (Wujastyk, 2003). John Parkinson, the king's herbalist, in his 1656 book *Paradise in Sole*, in the section titled "The Kitchen Garden", praised garlic as "The poor man's Treakle that is a remedy for all diseases. It is never eaten raw of any man that I know" (Parkinson, 1656).

Robert Thornton (1814) of Cambridge University, mentioned in *A Family Herbal*, says of garlic that when applied externally, it acts successively as a stimulant, rubefacient and blister. Internally, from its very powerful and diffusible stimulus, it is often useful in diseases of languid circulation and interrupted secretion. . . . In cases in which. . . . Irritability prevails, large doses of it may be very hurtful. . . . Taken in moderation, it promotes digestion; but in excess, it is apt to produce headach, flatulence, thirst, febrile heat and inflammatory diseases and sometimes a discharge of blood from the haemorrhoidal vessels. In fever of the thyphoid type, and even in the plague itself, its virtues have been much celebrated. . . . It is much recommended by some as an anthelmintic and has been frequently applied with success externally as a stimulant to indolent turmours, in cases of deafness. . . . and in retention of urine (Thornton, 1814).

Thornton, in his writings, gave several modes of application of garlic preperations:

> Several cloves may be taken at a time without inconvenience, or the cloves cut into slices may be swallowed without chewing. . . . The expressed juice, when given internally, must be rendered as palatable as possible, by addition of sugar and lemon juice. In deafness, cotton moistened with the juice is introduced within the ear and the application renewed five or six times in one day. Infusions of an ounce of bruised garlic in a pound of milk, was the mode in which Rosenstein exhibited it to children applicated with worms. Most commodious form for administering garlic is that of a pill or bolus conjoined with some powder. . . . Garlic made into an ointment with oils etc. and applied externally is said to resolve and discuss indolent tumours and has been by some greatly esteemed in cutaneous disease. It has likewise sometimes been emploved as a repellent. . . . Garlic applied to the soles of the feet. . . . in the confluent small pox, about the eight day, after the face begins to swell; and renewed once a day till all danger was over.
>
> **(Thornton, 1814)**

Garlic tea has been recommended for fever, headach, cholera and dysentery in China (Corzo-Martinez et al., 2007). In the Chinese pharmacopoeia, garlic is suggested for controlling parasites and ringworms, as well as for dysentery and gastero-intestinal disorders (Moyer, 1996). In India, garlic is used as an antiseptic lotion for washing wounds and ulcers (Corzo-Martinez et al., 2007).

Medicinal uses of garlic in the early 20th century are given in the *British Pharmaceutical Codex* and the *United States Dispensatory*. According to the *British Pharmaceutical Codex*:

> Garlic has antiseptic, diaphoretic, diuretic juice has been used in laryngeal tuberculosis, and poultices made from pulped garlic are useful in accessible tuberculosis lesions. The juice diluted with four parts of water has been used externally for treating suppurating wounds, and by inhalation for pulmonary tuberculosis.
>
> **(Pharmaceutical Society of Great Britain, 1934)**

According to the *United States Dispensatory*, garlic is used as a medicine. The garlic oil may be given in obstinate bronchitis and catarrhal pneumonia in young children. Burised garlic cloves are used in poultices for the lungs and were also applied to the feet for nervous restlessness or in the convulsions of young children (Hall, 1913).

In India, an ancient true story on garlic was written by Swami Satyadev Parivarjak (1954) and published by Kashi Nagrik Pracharini Sabha, Varanasi. The name of the book was *Lahsun Badshah (Garlic King).* In this book, Swami Satyadev recounts the true story of a citizen of Norway who was known as "Lahsun Badshah". He extensively traveled to different countries and promoted the medicinal uses of garlic among common people. When the Norwagian man meet Swami Satyadev, he was 150 years old and looked like a 35-year-old. He had rejuvenated his body twice with the help of garlic. He learned the method of bodily rejuvenation from the renouned scholar Swami Purnanand during his travels in the Himalayas. According to this book, the rejuvenation process requires three months.

1. **First month:**
   1. Three days of fasting with recitation of the Gyatri Mantra.
   2. On the fourth day, after three hours from morning to evening, five cloves of crushed garlic juice should be taken with 60 gm of butter for four days of the first week.
   3. In the second week, garlic clove juice along with 30 gm of honey should be taken.
   4. In the third week, ten raisins crushed with garlic cloves are taken four times.
   5. In the fourth week, five cloves of garlic crushed with one gooseberry are taken four times.

   RESULT: By following the earlier steps, the old skin of the body will rupture, and new skin tissue formation will start.

2. **Second month:**
   1. 250 gm of Indian cow milk are taken four times a day for one month.
   2. Garlic-mixed mustard oil is used daily for body massage for one month.

   RESULT: Old skin is replaced by new skin.

3. **Third month:**
   1. 200 gm of garlic cloves, 12 gm of honey, 12 gm of raisins without seeds and 36 gm of gooseberry should be mixed and taken four times a day for one month.

   RESULT: The whole body is rejuvenated in 91 days.

REFERENCES

Betz HH, Ed. (1992) *The Greek Magical Papyri in Translation.* University of Chicago Press, Chicago, IL.

Corzo-Martinez M, Corzo N, Villamiel M (2007) Biological properties of onions and garlic. *Trends Food Sci Technol.* 18: 609–625.

Gunther RT (1933) *The Greek Herbal of Dioscorides Illustrated by Byzantine, A.D. 512*. Edited and First Printed, Englished by John Goodyer. Dioscorides Pendanius of Anzarbos London, New York, Hafner, 1968.

Hall HO (1913) The onion (*Allium cepa*) and Garlic (*Allium sativum*) as a remedy for pneumonia and pulmonary tuberculosis. *Am Med*. 19: 26–34.

Manniche L (2006) *An Ancient Egyptian Herbal*. British Museum Press, London.

Moyers S (1996) *Garlic in Health, History and World Cuisine*. Suncoast Press, St. Petersburg, FL.

Nicholson PT, Shaw I, Eds. (2000) *Ancient Egyptian Materials and Technology*. Cambridge University Press, Cambridge.

Parkinson J (1656) *Paradise in Sole*. Richard Thrale, London.

Pharmaceutical Society of Great Britain (1934) *The British Pharmaceutical Codex*. The Pharmaceutical Press, London.

Rackham H (1971) *English Translation of Pliny the Elder's Natural History*. Harvard University Press, Cambridge, MA.

Swami Satyadev Parivarjak (1954). *Lahsun Badshah*. Kashi Nagrik Pracharini Sabha, Varanasi, pp. 1–111.

Thornton RJ (1814) *A Family Herbal*, 2nd ed. R and R Crosby, London.

Wujastyk D (2003) *The Root of Ayurveda: Selections from Sanskrit Medical*. Writings Penguin Books, London.

4 Health Benefits of Garlic

Scientific Research-Based Studies and Applications

Modern medical system recommendations for all drugs require intensive research to prove their efficacy. The FDA (Food and Drug Administration) provides guidelines for evaluating the scientific evidence against health claims. FDA requires the identification of a specific, measurable substance related to a disease or health condition. It also provides randomized, placebo-controlled, double-blinded intervention studies; prospective or retrospective observational studies; and disease risk determination (CFSAN, 2007). The final feature is publication in a peer-reviewed journal with a high impact factor. In this way, all possible effects of a drug—i.e. no effect, harmful effect or positive effect—are published to provide a balanced picture of the drug's impact.

The pharmacological activity of a garlic extract depends on the mode of its preparation. The extract prepared at room temperature contains mainly allicin, which has powerful antibacterial properties and a strong smell. Besides allicin, small amounts of several other thiosulfinates and complex sulphinyl components, including the antithrombotic ajoenes, are also present. The enzyme alliinase responsible for converting alliin to allicin is inactivated by heat. The water extract of heat-treated garlic mainly contains alliin, which is odor-free. Garlic powder is simply a dehydrated, pulverized garlic clove. The alliinase activity of garlic powder is identical to that of fresh garlic. It has been stated that the dehydration temperature should not exceed 60°C, above which alliinase is inactivated (Lawson, 1998). When distilled in steam, garlic yields an oily mass consisting of diallyl, methyl allyl and dimethyl, all originating from the thiosulfinates (Lawson et al., 1991; Lawson and Hughes, 1992). Another widely studied garlic preparation is aged garlic extract (AGE). When garlic is subjected to cold aging process, organosulfur compounds, such as S-allyl cysteine, s-allyl mercaptocysteine and several sulfur-containing amino acids, are produced.

Garlic is one of the most investigated medicinal plants. From 1960 to 2007, more than 3,000 research papers (Singh and Singh, 2008) have been published, and from 2002 to 2016, 2,836 papers were published on the chemistry and biological effects of garlic and garlic preperations (Sivasami, 2017). These studies were mainly focused on the cardiovascular, antimicrobial and anti-cancerous effects of garlic and, to a lesser extent, on the therapeutic indications for the treatment of hypoglycemia, heavy metal poisoning, liver dysfunction and hyperthyroidism (Lawson, 1993; Reuter and Sendl, 1995; Srivastava et al., 1995; Agarwal, 1996; Augusti, 1996; Block, 1998; Ross, 1999; Warber, 1999; Harris et al., 2001; Morbidone et al., 2001; Ichikawa et al., 2003; Tahiliani and Kar, 2003; Stephen, 2005; Ariga and Seki, 2006; Nies et al., 2006;

DOI: 10.1201/9781003646211-4

Mohamed et al., 2006; Eidi et al., 2006; Murugavel et al., 2007; Singh and Singh, 2008; Tesfaye and Mengesha, 2015; Alam et al., 2016; Singh et al., 2020; Batiha et al., 2020; Tellez et al., 2020; Tesfaye, 2021; Tadu et al., 2022; Bazaraliyaleva et al., 2022; Reda et al., 2024; Khedekar et al., 2024; Torres et al., 2024; El-Saadony et al., 2024; and Choudhury et al., 2024). They have extensivly reviewed the pharmacology and medicinal applications of garlic-isolated compounds for the treatment of a number of diseases (Table 4.1).

4.1 ANTIMICROBIAL ACTIVITIES

4.1.1 Antibacterial Activities

Louis Pasteur (1858) and Lehmann (1930) provided the first modern scientific evidence for the medicinal and antibacterial use of garlic extract. In early 20th century, garlic was scientifically recognized for its antibiotic properties (Reuter et al., 1996). Garlic vapors contain an active bactericidal component that is volatile at 37.5°C. There was no evidence of organisms developing resistance to this bactericidal substance. Boiled garlic and autoclaved garlic are non-germicidal (Walton et al., 1936).

TABLE 4.1
Active Constituents of Garlic and Their Pharmacological Actions.

Active component of garlic	Antibiotic	Antifungal	Hyperlipidemia	Antiplatelet aggregation	Anticancer	As a pesticide
alliin			+	+		
allicin	+	+	+	+	+	+
allixin					+	
adenosine				+	+	
Allyll,5-hexadienyl trisulfide				+	+	
Allyl methyl trisulfide				+	+	
s-allyl12-pro pene thiosul phinate				+		
ajoene	+	+		+	+	+
Diallyl disulfide	+		+	+	+	+
1,5-hexa dianyl trisulfide				+		
Methyl allyl trisulfide				+		
2-vinyl 1,3-dithiene				+		
3-vinyl 1,3-dithiene				+		
S-allyl merc ptocysteine					+	
Se-methylselenocysteine					+	
Allyl propyl disulfide			+			
Sodium 2-propenyl thiosulphinate				+		
s-methyl 1-cysteine sulfoxide			+			

Garlic has been observed to possess antiviral, antibacterial and antifungal activities (Cavallito and Bailey, 1944a, 1994b; Cavallito et al., 1944; Sharma et al., 1977; Yamada and Azuma, 1977; Amer et al., 1980; Lau et al., 1983; Tutakne et al., 1983; Caporaso et al., 1983; Elnima et al., 1983; Adetumbi et al., 1986; Yoshida et al., 1987; Srivastava et al., 1995; Harris et al., 2001; Lee et al., 2003; Kim et al., 2004; Shokrzadeh and Ebadi, 2006; Konaklieva and Plotkin, 2006; Dini et al., 2011; Karuppiah and Rajaram, 2012; Houshmand et al., 2013; Zain et al., 2013; Mikaili et al., 2013; Wallock-Richards et al., 2014; Blanch et al., 2016; Jang et al., 2018; Petropoulos et al., 2018; Shang et al., 2019). Cavallito and Bailey (1944a, 1944b) had demonstrated that garlic juice diluted to one part in 125,000 inhibits the bacterial growth of *Staphylococcus*, *Streptococcus* and *Vibiro* (including *Bacillus typhosus*, *B. dyseneterial* and *B. enteritidis*). Johnson and Vaughn (1969) reported that a 10% extract of dehydrated garlic bulb demonstrated antibacterial action against *Salmonella typhimurium* and *Escherichia coli* within two to six hours of exposure. Crude extract of garlic has proven to be quite effective against both gram-negative and gram-positive bacteria. Bacterial cultures resistant to the commonly used antibiotic chloramphenicol appear sensitive to garlic. Jezpwa et al. (1966), Adetumbi and Lau (1983), Khan et al. (1985), Hughes and Lawson (1991), Ahsan and Islam (1996), Tsao et al. (2003) and Zhou (2003) observed that the preparation of fresh garlic and vaccum-dried powdered garlic were effective against *Staphylococcus aureus*, *Streptococcus viridians*, *Streptococcus haematyticus*, *Klebsiella pneumoniae*, *Proteus vulgaris*, *Escherichia coli*, *Salmonella enteritidis*, *Bacillus subtilis*, *B. megaterium*, *B. Pumitus*, *B. mycoioles*, *B. thuringiensis*, *Sarcina lutea* and *Serratia marcescens*. Al-waili saloon et al. (2007) investigated that heating, storage and ultraviolet exposure of fresh garlic juice significantly inhibited its antimicrobial activity against common human pathogens. Hughes and Lawson (1991) noted that allicin, methyl allylthiosulfinates and allylmethyl sulfinates—constituents in aqueous garlic clove and powder homogenate—had *in vitro* antibacterial activity, while the polar compound alliin did not. Allicin and allyl-methyl plus methyl-allyl thiosulfinates have shown inhibition of the *in vitro* growth of *Helicobacter pylori*, the bacterium responsible for serious gastric diseases, such as ulcer and even gastric cancer (Canizares et al., 2004a, 2004b). Sasaki and Kita (2003) reported the antibacterial activity of garlic powder against *Bascillus anthracis*. A 1% water solution of garlic powder, using the test tube method, killed *B. anthracis* within three hours of treatment at room temperature.

Bakri and Douglas (2005) observed that garlic extract inhibits the growth of oral pathogens and certain proteases, and it may have therapeutic value, particularly for periodontitis. Periodontitis, a severe form of gun disease caused by oral bacteria, is commonly treated with antibiotics, such as amoxicillin, but antibiotic resistance is a growing problem. *Pophyromonas gingivitis* and other oral bacteria-causing periodontitis have been sensitive to allicin in fresh garlic extracts (Bakri and Douglas, 2005). This sensitivity is due to the very rapid reaction of allicin with the free thiol groups in bacterial enzymes cysteine proteases and alcohol dehydrogenases. These enzymes are essential for bacterial nutrition and metabolism, so the chance of resistance to allicin is a thousandfold less than that which occurs in the case of conventional antibiotics. The unpleasant taste and halitosis reported by the trial participants suggest that garlic mouthwash is unlikely to be a choice of consumer product (Groppo et al., 2007).

The antibacterial activity of garlic is mainly attributed to allicin. This is supported by the observation that if garlic extracts are stored at room temperature, the antibacterial effectiveness of garlic extracts is greatly reduced. This reduction was less if extracts are stored at 0–4°C. It indicates that the active component is thermally unstable. The intracellular effects of allicin are not well established. Allicin has a sulfydryl-modifying activity (Wills, 1956), and it is capable of inhibiting sulfydryl enzymes. Cysteine and glutathione counteract the thiolation activity of allicin, and in addition to the reaction mixture, the antibacterial activity is reduced. Garlic extract and allicin have been shown to exert bacteriostatic effects on some vancomycin-resistant enterococci. An inhibitory synergism was observed when used in combination with vancomycin (Jonkers et al., 1999). It is noticed that allicin modifies the sulfydryl groups on the enzymes of the Tn1546 transposon, which encodes vancomycin restance, enhancing susceptibility to vancomycin (Harris et al., 2001).

Helicobacter pylori are strongly linked to the development of chronic gastritis and gastric and duodenal ulcers. *H. pylori*–infected populations have an increased risk of stomach cancer. *H. pylori* has been found to be sensitive to various garlic-derived compounds, including allicin, ajoene, diallyl trisulfide and diallyl tetrasulfide, suggesting the possible use of one or more of these compounds for treatment (Sivam et al., 1997; Sivam, 2001; Ohta et al., 1999; O'Gara et al., 2000, 2008).

Garlic has long been recommended for the treatment of tuberculosis before the advent of streptomycin and other modern antibiotics. William C. Minchin of the Kells Union hospital in Ireland published a monograph on the utility of garlic preperations in the treatment of tuberculosis (Minchin, 1912, 1927). In his book, Minchin describes an inhaler mask containing a sponge soaked in garlic juice for the treatment of the disease. He also recommends the topical application of minced fresh garlic to diseased areas, mixed with hard and chewing raw garlic, and the use of fresh garlic spray for the throat in cases of laryngeal tuberculosis (Hall, 1913). The idea behind this method is that inhaled garlic juice would function as a volatile germicide and destroy the "tubercle spheres in the infected tissue of the human lung".

Torres et al. (2024) studied the antimicrobial activity of aqueous extracts of garlic and its derivaties for the design of pharmacological and industrial eco-friendly bactericidal products. A concentration of aqueous garlic extracts at 0.125 g/ml produced inhibition halos of 0.75 mm against *M. lutes*, 11 mm against *S. aureus* and 8.25 mm against *E. coli*. The bactericidal effect exerted by 1.0 g/ml of the extract was 210 min for *M. luteus*, 330 min for *S. typhi*, 270 min for *S. aureus* and no effect on *E. coli*. According to them, garlic extracts have effective antibiotic activity against foodborne pathogens that constantly infect humans. They also noted that garlic extract is more effective against Gram-positve bacteria compared with Gram-negative bacteria.

4.1.2 Antifungal Activities

The antifungal activity of garlic was first established in 1936 by Schmidt and Morquardt while working with epidermophyte cultures (Lemar et al., 2002). The antifungal activity of garlic extracts has been demonstrated in 40 species of zoopathogenic fungi, retarding the growth in 8 of the 15 genera tested (Appleton and Tansey, 1975; Shadkchan et al., 2004; Ledezma and Apitz-Castro, 2006). In a comparison of the fungistatic activity of garlic extract with nystatin, griseofulvin and amphotericin B, garlic showed broad-spectrum activity against 17 strains of fungi, including the dermatophytes, yeasts, Aspergillus and Penicillium. Garlic was more effective than nystain in retarding the growth of the fungi. Barone and Tansey (1977) and Yamada and Azuma (1977) demonstrated that allicin in aqueous extracts of garlic bulbs was the major anticandida constituent. Aqueous extracts of garlic are inhibitory and lethal to numerous strains of *Cryptococcus neoformans* (Fromtling and Bulmer, 1978). In China, garlic is frequently used to treat fatal forms of *Cryptococcae meningitis* (Abdullah et al., 1988). An aqueous extract of garlic has been demonstrated to inhibit the growth of several zoopathogenic fungi, such as *Candida albicans*, commonly involved in vaginitis; *Histoplasma capsulatum*, a fungus that produces a disease similar to tuberculosis; and fungi that cause athlete's foot and ringworm (Srivastava et al., 1995). Yoshida et al. (1987), in studying the antifungal activity of different purified fractions of garlic against *Aspergillus niger* and *Candida albicans*, observed that ajoene has stronger activity than allicin. Adetumbi et al. (1986) have reported that an aqueous extract of dehydarated garlic preparation inhibits the growth of the dimorphic fungus *Coccidioides immitis* and inhibits *in vitro* fungal spore germination. Adetumbi et al. (1986) and Lemar et al. (2002) reported that the reduction of *Candida albicans* growth by the aqueous garlic extract is due to the inhibition of nucleic acid, protein and lipid synthesis. Stephen (2005) reported that the breakdown products of allicin, the main parent antifungal compound in garlic, have a common mode of action against antimicrobial, anticancer and anticholestral properties. Some of the breakdown products of allicin cross cell membranes and combine with sulfur-containing molecular groups in amino acids and proteins, thus intefereing with cell metabolism. Human cells are not poisoned by allicin derivatives, thus preventing cell damage (Stephen, 2005). Allicin synergized the fungicidal activity of Cu^{2+} ions against various strains of fungi (Ogita et al., 2006). Allicin treatment caused Cu^{2+} solubilization and complexation with a plasma membrane protein.

Ajoene is an active compound in garlic that plays an antifungal activity against various fungi (Ledezma and Apitz-Castro, 2006). Sham-Ghahfarokhi et al. (2006) noted that garlic inhibits the growth of fungal infections as effectively as the drug ketoconazole when tested on *Malessezia furfur*, *Candida* species and 35 strains of various dermatophyte species. A report from a Chinese medical journal demonstrated the use of interavenous garlic to treat a potentially fatal and rare fungal infection of the brain called *Crytococcus meningitis*. In the report, the Chinese researchers compared the effectiveness of garlic with standard medical treatment, which involved a very toxic antibiotic called amphotericin B. The study revealed that intravenous garlic is more effective than the drug and was not toxic regardless of its dosage (Lemar et al., 2007). Sabitha et al. (2005) found that *Candida* colonies were substantially reduced in mice treated with garlic extract. Garlic stimulates phagocytic activity. This indicates that *Candida* infection is controlled via stimulation of the body's own defenses by garlic extract. Garlic oil can be used externally to treat ringworm, skin parasites and warts. Lesions caused by skin fungi in rabbits and guinea pigs were treated externally with garlic extract, and the wounds healed within seven days (Sabitha et al., 2005). Garlic exhibits antifungal activity against two species of airb orne pathogens: *Botrylis cierea* and *Trichoderma herzianum* (Lanzotti et al., 2012). Bakhshi et al. (2012) observed that the use of aqueous garlic is more effective than nystatin mouthwash to denature stomatitis.

Garlic essential oil suppresses the fungal pathogen *Penicillum funiculosum*, presumably by infiltrating membrane and compartments, distrupting the cytoskeleton and causing cytoplasmic and biomolecules leakage. In addition, proteomic analysis revealed the ability of garlic oil to increase/decrease the expression of few major proteins in relation to physiological metabolism (Li et al., 2014). The antimicrobial effect of Australian garlic extract was noticed against *Candida albicans*, *Listeria monocytogenes*, etc. Potent components of the garlic extract were identified using ultra-high-performance liquid chromatography and mass spectrometry (Phan et al., 2019). Pârvu et al. (2019) reported the antifungal activity of garlic extract against *Meyerozyma guilliermondii* and *Rhodotorula mucilaginosa* germination and growth. Another study reported the antifungal activity of various *A. sativum* extracts—namely, aqueous, ethanolic, methanolic and petroleum ether—against human pathogenic fungi, including *Trichophyton verrucosum*, *T. mentagrophytes*, *T. ruburn*, *Botrytis cinerea*, *Candida species*, *Epidermphyton floccosum*, *Aspergillus niger*, *A. glavus*, *Rhizopus stolonifera*, *Microsporum gypseum*, *M. audouinii*, *Alternaria alternate*, *Neofabraca alba* and *Penicillum expansum* (Fufa, 2019).

4.1.3 Antiviral Activities

Nagai (1973) reported *in vivo* antiviral effects of garlic in mice against intranasally inoculated influenza virus. The garlic extract also enhances the production of neutralizing antibody if inoculated with the influenza vaccine (Nagai, 1973). Investigation on the *in vitro* activity of garlic against selected viruses, including herpes simplex virus type 1 and type 2, parainfluenza virus type 3, vaccinia virus, vesicular stomatitis virus and human rhinovirus type 2, demonstrated that certain garlic constituents—(Viz.) ajoene, allicin, allyl methyl thiosulfinate, methyl allyl thiosulfinate—are potent virucides (Weber et al., 1992). Josling (2001), on the basis of his study on 146 volunteers, reported that an allicin-containing garlic supplement can prevent the infection of the common cold virus. Diallyl trisulfide in garlic bulb has antihuman cytomegalovirus (anti-HCMV) activity, and the mechanism is associated with suppression of IE gene transcription (Zhen et al., 2006).

Several laboratory tests have shown that garlic is effective against influenza B virus and herpes simplex virus. Two independent researchers in Japan and Romania have observed that garlic is able to protect living organisms from the influenza virus (Tasai et al., 1985). Tatarintsev et al. (1992) noted that ajoene acts in an HIV-infected cell system by inhibiting the integrin-dependent process. Allyl alcohol and diallyl disulfides are also effective against HIV-infected cells (Shoji et al., 1993). The antiviral activity of garlic extracts has been evaluated against influenza B, human rhinovirus type 2, human cytomegalovirus (HCMV), parainfluenza virus type 3, herpes simplex type 1 and type 2, vaccinia virus and vescular stomatitis virus. In *in vivo* experiment garlic extract showed protective activity against influenza viruses by improving the production of neutralizing antibodies in mice, which depends on ajoene, allicin, allyl methyl thiosulfinate and methyl allyl thiosulfinate (Sawai et al., 2008).

According to Bui et al. (2020), garlic essential oil is a powerful natural antiviral source that prevents coronavirus from invading the human body. Allyl disulfides and allyl trisulfides, which make up the majority of garlic essential oil, have the strongest anti-coronavirus action. Allicin-derived organosulfur compounds, such as ajoene, allitridin and diallyl sulfide, have been shown to have antiviral, immune-enhancing effects in *in vivo* and *in vitro* conditions (Rouf et al., 2020).

Garlic and its potent organosulfur components have been shown to promote antiviral activity by intracting with components on the exterior of the virion, which results in partial or complete inhibition of the virus entry step (Rouf et al., 2020). The aqueous extract of *A. sativum* expressed an effective inhibition of the viral infection of influenza, with a subtype of H1N1 showing an EC_{50} = 0.01 mg/ml in the Madin-Darby Canine Kidney (MDCK) cell line (Mehrbod et al., 2009). The exact value of EC_{50} (0.01 mg/ml) was also noted against measles morbillivirus in Vero cells treated with gold nanoparticles containing aqueous extracts of garlic. It causes obstruction of the viral receptors, stopping cell infection in the host cell (Melendez-Villanueva et al., 2019).

Quercetin, a flavonoid commonly present in fruits and vegetables, including garlic, possesses antiviral properties against the influenza virus and enterovirus by affecting viral attachments to the surface organosulfur compounds in garlic, such as allicin, ajoene and diallyl trisulfide, play an important role in the antiviral properties of garlic. These substances can prevent a virus from attaching to the host cell and affect viral RNA polymerase, which is required for viral replication. They can also prevent the viral process that changes the host cell's signaling pathway and prevent viral multiplication (Sharma, 2019). Even if the virus enters the host cell, there are still various steps suitable for its inhibition. One of these is the inhibition of the viral replication step. Replication may occur in the cytoplasm or in the nucleus of the host cell. Numerous investigations have proven the ability of *A. sativum* to prevent viral proliferation. The methanolic and ethanolic garlic extracts revealed the inhibition of viral RNA polymerase and nucleoprotein synthesis activity against the influenza A (H1N1) pdm09 virus. This property may be attributed to their ability to block viral attachment and suppress viral hemagglutinin (HA) (Chavan et al., 2016). Mohajer et al. (2016) studied the influence of aqueous garlic extracts on avian infectious bronchitis virus (IBV), a coronavirus that infects birds of the Coronaviridae family, and noted inhibition of the viral replication step. The retrovirus family utilizes the reverse transcriptase (RT) enzyme to convert viral RNA into DNA. In the treatment of retrovirus transmission, RT suppression is a major clinical strategy in the treatment of retrovirus transmission (Rouf et al., 2020). The hexane extracts of *A. sativum* are potent inhibitors of reverse transcriptase involved in human immunodeficiency virus I (HIV-1), with an IC_{50} of 64.08 ± 1.09 mg/ml (Silprasit et al., 2011). Relieve immune dysfunction is caused by the reticuloendothelosis virus (REV). Wang et al. (2017b) reported that allicin downregulated the ERK/MAPK signaling pathway, which caused the inactivation of REV replication.

Several plant products and their based drugs have been reported to have possible beneficial effects on COVID-19. *A. sativum* is one of them. Quercetin, a flavonoid present in garlic, bears the potential to inhibit the main protease of SAR-COV-2 (M^{Pro}), a crucial antiviral target (Khubber et al., 2020). Earlier, Chen et al. (2006) reported that quercetin-3-β-galactoside was a suppressor of SAR-COV-1 M^{Pro} with an IC50 of 42.79 ± 4.97 μm. There is a high structural similarity between the M^{Pro} of SARS-CoV-1 and SARS-CoV-2. In *silico* molecular docking of organosulfur compounds in garlic revealed an inhibitory potential of alliin against M^{Pro} of SARS-CoV-2. It was noted that alliin has potential antiviral activity against COVID-19 by binding to M^{Pro} of SARS-CoV-2 (Pandey et al., 2021). However, WHO declared that there is no confirmed indication that the use of garlic safeguards patients from the novel coronavirus. There is no doubt about the beneficial effects of garlic on human health. Although recent studies confirm that, despite garlic's inability to prevent COVID-19, its antiviral activity can play an important role in alleviating the suffering of patients with COVID-19.

4.1.4 Antiprotozonal Activities

Garlic is used against several disease caused by protozoan infection, such as *Entamoeba histolytica* (amoebic dysentery), *Girdia intestinalis* (giardiasis), *Plasmodium* species (malaria parasites), *Lesishmania* species (leishmaniasis), *Acanthamoeba castellanii* (Acanthamoeba keratitis) and *Trypanosoma brucei* (sleeping sickness).

Mirelman et al. (1987) reported that *Entamoeba histolytica*, which caused amoebiasis, is very sensitive to allicin, with total inhibition of *E. histolytica* observed at a concentration of 30 μg/ml. Allicin modifies the cysteine proteinase-SH group (Ankri et al., 1997). The LD_{50} of diallyl trisulfide against *E. histolytica* is 59 μg/ml, while the LD_{50} of diallyl trisulfide is 25 μg/ml, so tissue treatment is not recommended if levels exceed 25 μg/ml (Lun et al., 1994).

The parasitic protozoan *Giardia intestinalis* is the most commonly diagnosed cause of waterborne diarrhea. Several compounds found in garlic clove inhibit the growth of *G. intestinalis*. The 24-hour IC50 values for whole garlic extract, diallyl disulfide, 2-propenethial and allyl alcohol are 300 μg/ml, 100 μg/ml, 37 μg/ml and 7 μg/ml, respectively, as noted by Harris et al. (2000). The antigiardial activity of garlic compounds stimulates the production of nitric oxide synthase (NOS). Nitric oxide is cytotoxic to *G. intestinalis* (Harris et al., 2000).

Each year, more than 1,500,000 people die from malaria, a major infectious disease throughout the world. Malaria is transmitted by *Plasmodium* sporozites, which are injected into the skin of vertebrate hosts by an infected mosquito. Allicin causes inhibition of cysteine protease, which inhibits the processing of the *Plasmodium* circumsporozoite protein (CSP), thereby preventing the invasion of host cells both *in vitro* and *in vivo* conditions (Coppi et al., 2006).

Leishmaniasis is caused by intracellural parasitic haemoflagellates, which infest macrophages of the skin and viscera of the host. Garlic extracts are very toxic to *Leishmania major*, with an LD_{50} of 5 ppm (Khalid et al., 2005). Garlic-induced nitric oxide production may play a significant role in parasite killing (Gamboa et al., 2007). Ajoene, the major bioactive compound derived from garlic, showed a potent leishmanicidal activity *in vitro* against *Leishmania mexicana* and *L. donovani* (Ledezma et al., 2002). The 50% inhibitory concentration (IC50) for lysis was about 2 μM. They reported that the leishmanicidal activity of ajoene is due to the morphological alteration of the mitochondrial membrane and nuclear envelope, as well as the formation of large autophagic vacuoles. Gamboa-León et al. (2006) suggest that garlic extract has a mild protective effect against *Leishmania donovani*. This might be due to the nonspecific enhancement of interferon-gamma secretion.

Acanthamoeba castellanii is a free-living protozoan that causes the sight-threatening infection *A. keratitis*. Garlic extract components, diallyl polysulfides and allicin, are effective against this protozoan (Polat et al., 2008). *Tryponosoma* are unicellular, parasitic protozoa, which include *T. brucei*, the causative agent of sleeping sickness in humans and cattle disease. The LD_{50} of diallyl trisulfide against *T. brucei* is 0.8 μg/ml, which is significantly lower than the cytotoxicity of 25 μg/ml (Lun et al., 1994). Ajoene also exhibited antiparasitic activity by inhibiting the human glutathione reductase and *T. cruzi* trypanothione reductase (Gallwitz et al., 1999). Hazaa et al. (2016) reported that garlic oil kills *Plasmodium*, *Giardia*, *Leishmania* and *Trypanosoma* species.

Inhibitory activity of garli extract on *Giardia* was noted at 25 pg/ml, whereas the lethal dosage was 50 pg/ml. Soffar and Mokhtar (1991) demonstrated that garlic acts as an antigiardial agent and removed all symptoms from patients within 74 hours. It completely removed all indications of giardiasis from the stool within 72 hours at a dosage of 1 mg/ml aqueous extract twice a day. Allicin, ajoene and organosulfides from garlic are effective antiprotozonal compounds (Soffar and Mokhtar, 1991).

4.2 GARLIC AND CARDIOVASCULAR DISEASES

The medicinal effect of garlic and garlic extracts on cardiovascular diseases has been widely studied. Preperations of garlic and chemical constituents of garlic have been investigated for possible effects on cardiovascular diseases, such as hyperlipidemia, hypertension, platelet aggregation and blood fibrinolytic activity (Kendler, 1987; Lawson et al., 1992; Isensee et al., 1993; Isaacsolin et al., 1998; Koscielny et al., 1999; Mantawy and Mahmoud, 2002; Banerjee and Maulik, 2002; Gardner et al., 2003; Siegel et al., 2004; Rahman and Lowe, 2006; Omar et al., 2007; Espirito Santo et al., 2007; Gorinstein et al., 2007; El-Sabban and Abouazra, 2008; Maulik, 2013; Baliga et al., 2013; Kwak et al., 2014; Varshney and Budoff, 2016; Park et al., 2017; Takashima et al., 2017; Min et al., 2022).

Garlic-derived organosulfur compounds, such as diallyl disulfide and trisulfide and 2-propenethiol, inhibit cholesterol synthesis in cell cultures. They inhibit lanosterol 4α-methyl oxidase. Diallyl disulfide was active even at a level of 15 μmol/liter (Singh and Porter, 2006). Ajoene and allicin are inhibitors of cholesterol synthesis (Ferri et al., 2003). Gonen et al. (2005) suggested that allicin affects atherosclerosis not only by acting as an antioxidant but also by modifying lipoproteins and inhibiting low-density lipoprotein (LDL) uptake and destruction of macrophages. Although hydrogen sulfide (H_2S) and nitric oxide (NO) are metabolic poisons, they are considered important endogenously produced cell signaling molecules (gastrotransmitters) with key roles in cardiovascular processes, i.e. dilation of blood vessels, hypertension and myocardial infarction (Chang and Chen, 2005; Aquilano et al., 2007).

4.2.1 Hypocholesterolaemic Action

Several studies related to the effects of garlic on blood lipids in patients with cardiovascular disorders have been carried out in India. Intake of high-fat meals causes a significant increase in serum triglyceride and cholesterol levels (Groot and Scheck, 1984). Use of garlic/onion prevents the hypercholesterolemia induced by high-fat meals (Jain and Andleigh, 1969; Bordia and Bansal, 1973; Sainani et al., 1979; Mirhadi and Singh, 1991; Elkayam et al., 2003). Augusti (1977) and Bhushan et al. (1979) reported that an intake of 10 g of fresh garlic per day for two months significantly decreases (15%–28.5%) serum cholesterol levels among hypercholesterolemic patient. Sainani et al. (1979) compared vegetarians with different eating habits with regards to the use of onions and garlic. Group-I consumed onion and garlic in libral amounts (50 g garlic and 600 g onion per week). Group-II consumed a small amount of onion and garlic (10 g garlic and 200 g onion per week). Group-III totally abstained from using garlic and onion. The levels of cholesterol, triglyceride, phospholipid and β-lipoproteins were the lowest in individuals consuming liberal amounts of garlic and onion. These results indicate that regular consumption of onion and garlic in the diet has a beneficial effect on maintaining serum lipid levels at low or normal levels.

Total serum cholesterol is an important factor in the development of ischemic heart disease (IHD). Cholesterol present in the β-lipoprotein (LDL, low-density lipoprotein) and pre-β-lipoprotein (VLDL, very low-density lipoprotein) fractions finds its way into the aterial wall, whereas α-lipoprotein (HDL, high-density lipoprotein) cholesterol helps to reduce the serum cholesterol level. An increase in the HDL/LDL ratio has a preventive effect on the development of IHD. Jain (1977) and Bordia (1981) observed that consumption of 0.25 mg/kg garlic oil caused a steady decrease in the LDL + VLDL levels, with a concomitant increase in HDL levels. Ernst et al. (1985), Orekhov et al. (1995), Efendy et al. (1997) and Ide et al. (1997) have noticed that the aged garlic extract, aqueous extract from garlic powder and its constituents lower serum cholesterol levels and inhibit the oxidative modification of LDL, thus presumed to have a protective effect against atherosclerosis.

Gebhardt (1993) reported the multiple inhibitory effects of garlic extracts at different steps in the cholesterol biosynthesis pathway in human Hep G_2 cells. According to him, defined compounds (allicin) present in water-soluble extracts of garlic inhibit the biosynthesis of cholesterol in hepatocytes, thus contributing to the reduction of serum cholesterol. Garlic-derived components are capable of combining with the sulfhydryl (–SH) group (Augusti and Mathew, 1974; Block, 1985). Reduced conversion of acetate into cholesterol has been observed both *in vivo* and *in vitro* (Chang and Johnson, 1980; Saxena et al., 1980). AS-SH groups are involved in several metabolic pathways. It may be inferred that garlic, besides inhibiting lipid synthesis (Eilat et al., 1996), would affect other metabolic activities as well (Srivastava et al., 1995). Larner (1995) proposed the tellurium hypothesis of the hypocholesterolaemic action of garlic. According to him, tellurium compounds found in high concentration in fresh garlic bulb may contribute to hypocholeserolaemic action by inhibiting squalene expoxidase, the penultimate enzyme in the synthetic pathway of cholesterol. The oxidation of LDL plays an important role in the initiation and progression of atherosclerosis. Yamaji et al. (2004) suggested that the use of egg-yolk-enriched garlic powder (EGP) has an antiatherogenic effect. EGP inhibits copper-induced LDL oxidation in a dose-dependent manner. The use of EGP significantly suppressed the production of peroxides in 60 cells and protects endothelial cells from hydrogen peroxide–induced cell injury. Sumioka et al. (2006) reported that monascus garlic fermented extract (MGEE) attenuates hyperlipidemia, suggesting that it is a potent agent for preventing arteriosclerotic diseases. Intake of enteric-coated garlic powder (equal to 400 μg, 1 mg allicin) twice daily has significantly reduced total cholesterol, LDL-cholesterol and triglyceride, while increasing HDL-cholesterol (Kojuri et al., 2007).

Garlic inhibited liver lipogenesis and cholesterol-forming enzymes, such as malic acid synthetase, fatty acid synthetase, glucose-6-phosphate dehydrogenase and 3-hydroxy-3-methylglutaryl-CoA (HMG-CoA) reductase (Lau, 2006). According to him, water-soluble organosulfur compounds, such as S-allyl cysteine (SAC) found in mature garlic extract and diallyl disulfide (DADS) found in garlic oil, are effective inhibitors of cholesterol formation in *in vitro* studies. Low-density lipoproteins (LDL) are suppressed, and LDL resistance to oxidation is increased (Lau, 2006). A meta-analysis including 39 primary trails on the effects of two months of garlic preparation administration on total cholesterol, low-density lipoprotein cholesterol, high-density lipoprotein cholesterol and triglycerides was performed by Ried et al. (2013b). In these human studies, it was noted that garlic preparation significantly decreased the serum cholesterol and triglyceride (Ziae et al., 2001; Ried et al., 2013b).

Garlic reduces cholesterol level either by stimulating the acidic and neutral steroids excretion or by reducing the cholesterogenic and lipogenic effects of fatty acid synthase, 3 hydroxy-3-methyl-glutaryl-CoA reductase, malic acid and glucose-6-phosphate dehydrogenase in hepatocytes (Qidwai and Ashfaq, 2013). Garlic was found to have an important effect on dyslipidemia by significantly decreased serum TC, TG and LDL levels, while moderately elevating HDL cholesterol (Iweala et al., 2005). They reported that ethanolic garlic extract uptake to albino rabbits caused decreased cholesterol level and body weight. In clinical trials, Sobenin et al. (2008, 2010) demonstrated that garlic administration at doses of 300 mg/day and 60 mg/day for 12 months and 12 weeks, respectively, decreased TC, TG and LDL, with an increase in HDL. Garlic has also been shown to lower serum cholesterol (Ried, 2016). Garlic flour formulations were found to be a universal cardiac and circulatory tonic, lowering blood TC, LDL-C and HDL-C (Kwak et al., 2014).

Shabani et al. (2018) noted that garlic reduces lipid profile as well as glucose parameters and is therapeutically effective in patients suffering from cardiovascular diseases and diabetes. Zhenyne et al. (2023) noted that, compared to placebo, garlic significantly reduced the lipid profile. Meta-regression and subgroup analysis showed that among six covariates—total sample size, the mean age, the mean dose, the duration of intervention, the oral form of garlic and dietary intervention—age and sample size are potential effect modifiers.

Ajamzibad et al. (2021) found that boiled garlic and shirazi lemon significantly reduced the blood lipid profile. Li et al. (2022) conducted a clinical trial showing that even a small quantity of regular garlic used by cardiovascular disease patients caused a significant reduction in lipid profile. Shahzad et al. (2023) also suggested that consuming raw garlic has a positive effect on lowering the lipid profiles in patients with hyperlipidemia. A significant reduction in blood cholesterol was noted after consuming raw garlic for 42 days.

4.2.2 Antihypertensive Activity of Garlic

Hypertension is a well-known risk factor for cardiovascular mortality, affecting about 1 billion individuals all over the world (Ried et al., 2008). High blood pressure is one of the major risk factors of atherosclerosis (Srivastava et al., 1995). The hypertensive effect of garlic was recognized by Loeper and De Bray (1921). Damrau (1941) gave two Allimin tablets containing 4.75 g of garlic concentrate (equivalent to about 0.31 g of desiccated garlic and 2.375 g of desiccated parsley) to 26 hypertensive patients three times daily for three days. Blood pressure reduction was noted in 85% of the patients. The average decline in systolic and diastolic blood pressure was 12.3 mmHg and 6.5 mmHg, respectively. He also reported that headache was relieved in 14 of the 17 patients complaining of the symptom, dizziness was cured in 12 of 13 patients, and significant improvement was observed in the remaining patients.

Garlic has been used against hypertension in China and Japan and is recognized officially for this treatment by the Japanese food and drug administration (Bolton et al., 1982). Piotrowski (1948), Papayannopoulos (1969), Srinivasan (1969) and Anonymous (1986) have also demonstrated garlic's antihypertensive effect. Pectov (1979) cited several studies on the hypotensive effect of garlic from the Soviet Union and Bulgaria. Aqel et al. (1991) suggested that hypotensive action of garlic juice may be due to a direct relaxant effect on smooth muscles. Although components of garlic do exert pharmacological and therapeutic effects on hypertension (Srivastava et al., 1995), the nature of the garlic component responsible is not known. Andrianova et al. (2002) observed that allicor treatment of 600 mg/day caused a reduction in both systolic and diastolic blood pressure.

Meta analysis done in mid-2008 suggest that garlic is associated with blood pressure reductions in patients with an elevated systolic blood pressure (SBP). Garlic reduced the SBP by an average of 8.4 mmHg by Ried et al. (2008) and 16.3 mmHg by Reinhart et al. (2008). According to Ried et al. (2008), the effects of garlic on blood pressure are comparable to the hypotensive effects of commonly prescribed blood pressure drugs. The blood pressure–reducing properties of garlic is associated with its hydrogen sulfide–producing ability (Benavides et al., 2007).

Red blood cells (RBC) convert diallyl disulfide and trisulfide to hydrogen sulfide (H_2S) molecules, which occupy the membranes of the RBC. H_2S is normally a toxic, foul-smelling gas when produced in the body in minute amounts through the enzymatic breakdown of cysteine; however, it plays a vital role in helping cells communicate with each other (Jacob et al., 2008). The formation of H_2S from freshly crushed garlic was demonstrated when garlic juice was mixed with a suspension of human red blood cells, such that the amount of garlic juice in the RBC suspension was equivalent to two cloves of garlic in the blood volume of a typical adult human (5L). With the help of a very sensitive polargraphic H_2S sensor, it was noted that substantial production of H_2S immediately occurred mainly at the surface membrane of the RBC (Benavides et al., 2007; Lefer, 2007). Furthermore, Chuah et al. (2007) reported that garlic contains a compound S-allyl-L-cysteine that boosts the production and circulation of H_2S in the body. Hydrogen sulfide plays a key role in regulating blood pressure like nitric oxide, releasing blood vessels by activating an enzyme called guanylyl cyclase located in the vessel's smooth muscle cells. Recently, scientists have determined that H_2S has the same dilating effect on the vessels as nitrous oxide (NO), but it acts through a completely different pathway. H_2S gas is generated by the smooth muscle cells and endothelial cells that make up blood vessels. H_2S gas activates K_{ATP} channel proteins in the smooth muscle cell membrane, prompting potassium ions to exit the cell. Electric current created by the flow of potassium ions out of the cell limits the flow of calcium ions into the cell. This change relaxes the cell and dilates the blood flow. Hydrogen sulfide from garlic (fresh is better than processed garlic product) has been shown to help prevent the severe damage that occurs in heart attacks (Chuah et al., 2007; Elrod et al., 2007; Mukherjee et al., 2009) (Figure 4.1).

Asdaq and Inamdar (2010) reported that concomitant use of fresh garlic homogenate (FGH) or its bioactive constituent S-allyl cysteine sulphoxide (SACS) with captopril is very effective in controlling high blood pressure. The combined therapy of fresh garlic homogenate (FGH) 250 mg/kg with captopril (CAP) caused an effective reduction of sustolic blood pressure (SBP). The combined therapy of FGH with CAP resulted in a significant fall in LDH and creatine kinase myocardial band (CK-MB) activities in serum and an elevation in heart tissue homogenate. This combined therapy exerted a synergistic effect on blood pressure reduction and inhibition of angiotensin-converting enzyme (ACE).

Several clinical studies shared that the use of garlic reduced blood pressure in more than 80% of patients suffering from high blood pressure (Stabler et al.,

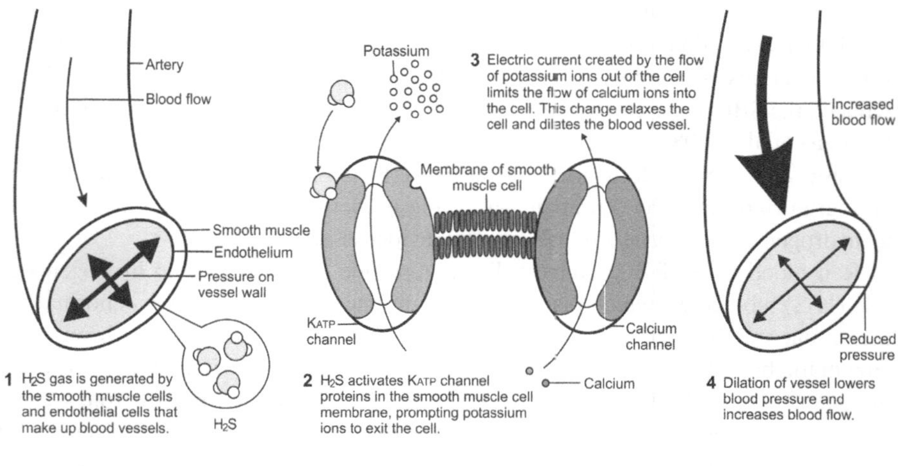

FIGURE 4.1 Sulfide-containing compounds found in garlic are converted into H_2S by molecules that occupy the membranes of the RBC. Furthermore, garlic contains a compound called S-allyl-L-Cysteine that boosts the production and circulation of H_2S in the body. Hydrogen sulfide plays a key role in regulating blood pressure by dilating the blood vessels. The dilating effect of H_2S on blood vessels lowers blood pressure like nitric oxide, but it acts through a different pathway.

2012; Omar, 2013). They reported that aged garlic extract was superior to placebo in lowering systolic blood pressure in patients suffering from uncontrolled hypertension. A dosage of 240–960 mg of aged garlic extract containing 0.6–2.4 mg of S-allyl cysteine significantly lowered blood pressure in 12 weeks (Ried et al., 2013a). Barankovic et al. (2011) noted that intravenous administration of garlic extract produced dose-dependent, reversible hypotensive and brodycardic effects in anesthetized normotensive rates. Asdaq and Inamdar (2010) showed the protective effects of combined therapy with garlic homogenate and hydrochlorothiazide (HCTZ) in animals with hypertension and myocardial damage. Garlic combined with HLTZ increases the lactate dehydrogenase, creatinine phosphokinase and superoxide dismeustage and catalase activities in the heart, whether used concomitantly or separately. There was a significant restoration of body weight, systolic blood pressure and lipid profile in all groups. A moderate dose of garlic was more effective than a low dose. The moderate dose of garlic with HCTZ has synergistic cardioprotective and antihypertensive properties against fructose- and isoproterenol-induced toxicities (Asdaq and Inamdar, 2011a).

Different experimental and human studies reported the antihypertensive effect of garlic extracts and its derived bioactive molecules. Sobenin et al. (2012) showed the plasma fibrinolytic activity of garlic extracts, and they found that it increased fibrinolytic activity in both healthy and acute myocardial infarction participants. *In vivo* experiments exhibited the antihypertensive effect of aqueous garlic extract in the "2 Kidney 1-clip" model of hypertension in rats by reducing thromboxane B2 and prostaglandin E_2 levels, which reduced hypertension in tested rats (Dubey et al., 2017). The use of garlic at a dose of 100 mg/kg for five days completely prevented acute hypoxic pulmonary vasoconstriction caused by endothelin-1 and angiotensin II production (Drobiova et al., 2011). The mechanism of the antihypertensive effect of garlic extracts is due to the presence of many active sulfur molecules that have been involved in stimulating endothelium-constructing and relaxing factors, leading to lower blood pressure. Garlic has also been shown to stimulate the production of both nitric oxide (NO) and hydrogen sulfide (H_2S), which ultimately leads to vasodilation and a reduction in high blood pressure (Ried and Fakler, 2014).

4.2.3 Antiplatelet Aggregation Activity

Platelets are small particles, which circulate in the blood of mammals. Human blood contains 150,000–400,000 platelets per microliter. They play an important role in cardiovascular hemostasis. The major function of the blood platelets is to maintain the hemostatic integrity of the blood vessel and to stop bleeding after injury. Platelets interact with the subendothelial components of the vessel wall, such as collagen, that activate the platelets in which it brings about a series of changes ranging from change of shape, production of pseudopodia and adherence to collagen fibrils. An endothelial injury of the blood vessels leads to the exposure of subendothelial collagen to circulating blood cells and the accumulation of macrophages and platelets at the site of injury. In this process, platelets secrete many chemicals, including vasoactive substances and platelet-derived growth factor (PGDF), the latter of which induces the proliferation and migration of smooth muscle cells, resulting in the formation of a proliferactive atherosclerotic lesion. Evidence that platelets directly enhance cholesterol esterification and macrophage cholesterol ester accumulation further underscores the important role platelets play in the development of athrosclerosis (Curtiss et al., 1987; Srivastava and Mustafa, 1993). 95% of all cases of coronary heart disease (CHD) are due to coronary atherosclerosis. During plaque formation, changes in hemodynomics increase plaque instability, which can easily lead to plaque rapture. Once the plaque raptures, platelet activation in the blood, activation of the coagulation cascade and multicellular mobilization lead to platelet-rich emboli that block blood vessels and cause malignant clinical events, such as ischemic myocordial infraction and stroke (Kobiyama and Ley, 2018).

Garlic has been extensively studied for its effects in preventing atherosclerotic disease. Reuter et al. (1974) and Lutomski (1987) demonstrated that adenosine, a compound in high concentration (0.056%) in garlic, inhibits the aggregation of platelets and improves blood flow in the coronary vessels. Investigations have shown that garlic oil could inhibit platelet aggregation (Mohammad et al., 1980; Mohammad and Woodward, 1986; Srivastava and Justesen, 1989). Ether extract of garlic juice taken with a fatty diet could decrease cholesterol and fibrinogen and increase fibrinolytic activity and blood coagulation time. Garlic can lead to a decrease in thrombocyte aggregation in blood (Bordia et al., 1977; Srivastava, 1984; Harenberg et al., 1988; Jung et al., 1989; Kiesewetter et al., 1991; Lawson, 1993; Srivastava and Tyagi, 1993; MacDonald et al., 2004). Srivastava and Mustafa (1993) reported that several mechanisms are involved in garlic-induced inhibition of platelet aggregation. These effects are the modification of platelet membrane properties, inhibition of calcium mobilization and inhibition of several steps of the arachidonic acid cascade in platelets. In fact, a direct inhibitory effect of garlic extracts and its components on the enzymes of archodonic acid cascade has been reported (Srivastava, 1984; Srivastava and Justesen, 1989; Makheja and Bailey, 1990; Wagner et al., 1991).

Day et al. (1976) reported that several biochemical agents, such as adenosine diphosphate (ADP), collagen, epinephrine, arachidonate and especially thromboxane A, elevate platelet aggregation. Bordia et al. (1978) observed that garlic caused a dose-dependent inhibition of platelet aggregation induced by ADP, epinephrine and collagen. Both aqueous and organic extracts of garlic inhibit platelet aggregation induced by arachidonic acid, ADP, adrenaline, collagen, calcium inophore A23187 and thrombin (Apitz-Castro et al., 1983; Srivastava, 1984; Srivastava and Justesen, 1989).

Makheja and Bailey (1990) reported that adenosine is one of the most important ingredients of the garlic bulb. About half of the antiplatelet activity of garlic has been attributed to adenosine. They also reported that adenosine and allicin present in garlic inhibited platelet aggregation without altering the arachidonic acid (AA) metabolism in platelets. Lawson et al. (1992) reported the antiaggregatory activity of different galic preperations. In platelet-rich plasma, most of the antiaggregatory activity of the garlic clove homogenates were due to adenosine; however, in whole blood, the antiaggregatory activity was due to allicin and other thiosulfinates. Garlic powder tablets were equally as active as clove homogenates, whereas steam-distilled oils and oil macerates were only 35% and 12% active against antiaggeregatory activity, respectively. In steam-distilled, oils most of the activity was due to diallyl trisulfide. In the oil macerates, most of the activity was largely due to the vinylditheins. Polysulfides, such as diallyl trisulfide (Srivastava and Mustafa, 1993) and dimethyl trisulfide (Makheja and Bailey, 1990), are shown to be antiaggregatory and alter AA metabolism. The inhibition of platelet aggregation ultimately decreases atherosclerosis and thrombosis (Mirhadi and Singh, 1991). Kiesewetter et al. (1991) observed that the intake of 800 mg of garlic powder over a period of four weeks caused the disappearance of thrombocyte aggregation, increased microcirculation of the skin by 47.6% and decreased plasma viscosity by 3.2%; decrease in diastolic blood pressure by 9.5% and blood glucose concentration by 11.8% were also observed. Jung et al. (1991) reported that five hours after the intake of 900 mg garlic powder caused a significant increase in capillary skin perfusion by 5.5%. It causes minimal vasodilation and significant reduction in haemotocrit and plasma viscosity. Apitz-Castro et al. (1983) reported that the methanol extract of garlic and some pure fractions inhibited platelet aggregation. The antiplatelet property of garlic is especially due to one of its components, (E)- and (Z)-ajoene, which perturbs the membrane by lodging itself deep between the two monolayers of the plasma membrane of intact platelets. Jain and Apitz-Castro (1987) reported that ajoene did not alter AA metabolism in intact platelets. It is of interest to note that ajoene was shown to synergistically potentiate the antiaggregatory effects of prostacyclin, forskolin, indomethacin and dipyridamole (Apitz-Castro et al., 1986). Chang et al. (2004) identified sodium 2-propanyl thiosufinate (2PTS) from boiled garlic, which significantly inhibits ADP-induced platelet aggregation at 0.1 mM in human platelets. In contrast, the maxial aggregation percentage returned to normal level at 1 mM of alkenyl thiosulfinate in human platelets, suggesting that 2PTS has the potential to promote immune function and prevents cardiovascular disease. Cavagnaro et al. (2007) suggest that processing of garlic and the conditions used for cooking can markedly influence its effectiveness as a platelet inhibitor. They are of the opinion that crushing garlic before moderate cooking can reduce the loss of antiplatelet activity.

Lawson (2007), in his long-term study on nine people given 8 grams of raw garlic per day (35 mg allicin content) or an equivalent amount of boiled garlic (84 mg allicin content; no allicin), both in the form of sandwiches, observed the following: after four weeks, while boiled garlic consumption had no effect, raw garlic consumption decreased the platelet-rich plasma (PRP) aggregation by about 5–8% in all nine people. Simultaneously, it was increased by 10–13% in all cases. It indicates that consumption of normal levels of raw or boiled garlic (1–3 g) will have an effect on *ex vivo* aggregation in healthy individuals.

Teranishi et al. (2003), in *in vitro* testing of baboon blood-derived platelet-rich plasma, (PRP) showed that platelet aggregation with ADP (200 μM or 40 μM) was completely inhibited in a dose-dependent manner by ajoene at 75 μg/ml (320 μM). Aggregation with collagen (12.5 μg/ml) was inhibited from 66% to 11% by ajoene at 25 μg/ml (101 μM), but higher concentrations had little additional effect (Teranishi et al., 2003). *In vivo* testing with baboons showed that ajoene in soyabean oil at a dose of 25 mg/kg, given intravenously either once or repeatedly at intervals of two to three hours, completely inhibited platelet aggregation.

In vitro study demonstrated that allicin and propyl propanethosulfinate (Pr S(O) S Pr) rapidely interact with –SH groups on intracellular cysteine proteins, especially calpain (calicum-dependent enzyme), both at the platelet surface and its interior. They inhibit aggregation at 10 to 20 μM levels (Badol et al., 2007). The small molecule methyl methane thiosulfinate (Me S(O) S Me) enters the platelet more rapidly than allicin and reacts mainly inside the cell, without affecting the membrane surface (Badol et al., 2007). Thiosulfinates and diallyl trisulfides are potent inhibitors of calcium ion mobilization, which also accounts for their ability to inhibit platelet aggregation (Qi et al., 2000). Hiyasat et al. (2009) reported that an alcoholic extract of *A. sativum* caused significant antiaggregatory activity towards human platelets. Allison et al. (2012) observed that aged garlic extract inhibited the binding of ADP-activated platelets to immobilized fibrinogen.

Garlic has been showen to inhibit platelet aggregation in both *in vitro* and *in vivo* experiments (Chakraborty and Majumder, 2020). Garlic antithrombotic impact is noted in *in situ* studies by modifying fibrinolytic activity through increased plasminogen activity and inhibition of thrombin production (Fukao et al., 2007). Chakraborty and Majumder (2020) have demonstrated that garlic suppliments reduced the cyclogenase activity of thromboxane A2 production, resulting in anti-platelet action. Irfan et al. (2018) reported that fermented preparations of garlic and chitosan (a natural polysaccharide obtained from chitin) possess stronger antiplatelet effects than their non-fermented counterparts. According to them, this is due to the increased bioavailability of active compounds produced during fermentation. Lia et al. (2021) tested aqueous extracts of garlic (AEG) and black garlic (AEBG) (a product of garlic, which tastes delicious and does not smell) in ten groups as vehicle control, positive controls, as well as the concentration groups (each 1–4 μg/ml) to observe their antiplatelet activity in blood. The outcome of the experiment showed that both the preperations—i.e. AEG and AEBG—casued significant inhibition of platelet aggregation in human blood.

TXB2 and prostaglandin 12 (PG12) are metabolites of arachidonic acid (AA). Thromboxane A2 (TXA2) induces vasoconstriction and platelet aggregation and promotes arachidonic acid formation, whereas PG12 functions in vasodilation and inhibition of platelet aggregation. TXA2 and PG12 are usually determined by measuring their stable metabolites, TXB2 and 6-keto prostaglandin F1α (6-keto-PGFα). Total saponin of garlic (TSG) reduces the level of TXB2 and increases the level of 6-keto-PGF1α, maintaining the ratio of TXB2 to 6-keto-PGF1α in a relatively stable dynamic equilibrium to maintain vascular homeostasis (Miao et al., 2020). Adhesion molecules such as vascular cell adhesion molecule-1 (VCAM-1) and intracellular cell adhesion molecule-1 (ICAM-1) promote platelet adhesion and leukocyte recruitment and play an essential role in arachidonic acid formation. Garlic powder decreased VCAM-1 and ICAM-1 level, significantly increasing activated partial thromboplasting time (APTT). It suggests that garlic might contribute to the treatment of AS by dealaying clotting time, altering angiotensin and decreasing the expression of VCAM-1 and ICAM-1.

Tissue factor (TF) is important in initiating the extrinsic coagulation cascade pathway, which is responsible for producing thrombin from prothrombin via activation of factor VII (Dahlback, 2000). Subendothelial TF is also responsible for initiating fibrin formation at sites of vascular injury, and blood-borne TF may be a vital contributor to the propagation of the developing thrombus (Grover and Mackman, 2020). It has been reported that tumor necrosis factor-α (TNF-α)-induced TF mRNA expression in human umbilical vein endothelial cells (HUVECs) was suppressed by inhibition of JNK. Therefore, inhibition of the JNK pathway by diallyl trisulfide (DATS) may inhibit the induction of TF by TNF-α. DATS inhibited not only TF activity but also TF mRNA and protein expression *in vitro* (Okue et al., 2022). Garlic is a promising food and an alternative or complementary therapy to antiplatelet therapy.

4.3 ANTIDIABATIC EFFECTS OF GARLIC

Diabetes mellitus is one of fastest growing metabolic illnesses of the current global population. The prevalence of diabetes mellitus in adult humans was about 153 million in 1980 and 347 million in 2008 (Danaei et al., 2011). According to Shaw et al. (2010), it becomes 439 million in 2030, with a 69% increase in developing countries and a 20% in developed countries. Diabetes is a condition defined by hyperglycemia, which impaires the functionality of numerous organs and tissues in affected individuals. In diabetes, hyperglycemia is caused by inadequate insulin secretion from pancreatic β-cells in response to glucose stimulation, as well as insulin's ineffectiveness in regulating the metabolism of its peripheral target tissue. Type 1 diabeties is caused by the autoimmune destruction of the pancreatic β-cells, which leads to a complete lack of insulin formation, and as a result, insulin resistance (Yki-jarvienen et al., 1986; Greenbaum, 2002) develop. Type 2 diabetes mellitus (T2DM) typically occurred in overweight young adults as well as aging population (Semenkovich, 2015).

According to Liu et al. (2007), though promising results have been obtained from animal studies on the use of garlic as an antidiabetic agent, the antidiabetic effects of garlic is still in the early stages. Extensive clinical investigations are needed to establish the antidabetic efficacy of garlic, as its antidiabetic potential has gained much attention over the last two decades (Ashraf et al., 2005; Atkin et al., 2016; Ryu and Kang, 2017; Wang et al., 2017a; Choudhary et al., 2018; Ghorbani et al., 2019; Parham et al., 2020; Kumar et al., 2021; Sanie-Jahromi et al., 2023). Most of the investigations on the antidiabetic effect of garlic are done on blood glucose control in healthy animals, animal models and diabetic patients (Saikat et al., 2022; Sanie-Jahromi et al., 2023). Diabetes is usually induced in animal models with the help of alloxan or streptozotocin. Liu et al. (2006), in his study, found that garlic oil and diallyl disulfide did not have antidiabetic effects immediately. It took a few weeks for the results to appear. Afkhami-Ardekani et al. (2006) noted that oral intake of 900 mg garlic, twice a day, did not reduce fasting blood sugar or two-hour postprandial glucose levels in type 2 diabetes patients. Dafriani et al. (2020) also reported that the administration of thre pieces of garlic combining with antidiabetic drugs for two weeks has not been able to reduce the blood sugar levels to normal limits. A combination of allicin, found in intact garlic cloves, with antidabetic drugs like glibenclamide and insulin has shown moderate effectiveness in regulating hyperglycemia compared to the use of antidabetic drugs alone (Sheela et al., 1995). In diabetic mice, garlic extract reduced serum glucose, total cholesterol, triglycerides, urea, uric acid, aspartate aminotransferase and alanine aminotransferase, while significantly increasing the serum insulin levels (Eidi et al., 2006). The volatile sulfur components in garlic are primarily responsible for its antidiabetic action with insulin resistance management (Pandiya and Banerjee, 2013). Takim et al. (2021) reported that garlic is effective in the treatment of diabetes by inhancing the gene expression of caspase 3 and caspase 9. Zhai et al. (2018) found that garlic has a nutrional role in preventing diabetes. According to them, alliin modulates the intestinal microbe composition by reducing lachnospiraceae and augmented ruminococcaceal in diet-induced obese mice. According to Sujithra et al. (2018) and Goudappeala et al. (2021) the hypoglycemic effect of garlic is due to the presence of allylepropyl disulfide or diallyl disulfide and their effect on purine metabolism. Garlic oil and diallyl trisulfide combination treatment in rats increased the proportion of glucose-to-glycogen conversion (Gouddappala et al., 2019). Swanston et al. (1990) showed that the blood glucose-regulating activity of garlic extract was due to the presence of sulfur-containg compounds and flavonoids. According to Kasuga et al. (1999), garlic was effective in regulating the corticosterone levels in adrenal hypertrophy, which may ultimately help prevent an increase in blood glucose in diabetic mice.

4.3.1 Garlic as Insulin Secretagogues

Garlic has insulin secretagogue activity, though the detailed mechanism is not exactly known (Mohana et al., 2012). Use of 100, 300 and 600 mg/kg aged garlic extract resulted in high serum insulin levels in diabetic rats (Eidi et al., 2006; Thomson et al., 2007). Liu et al. (2005) found that in diabetic rats fed with garlic oil or garlic extract, insulin levels were significantly increased. Shaker et al. (2022) noted that garlic extract may increase pancreatic beta cell function by upregulating the peroxisome proliferator-activated receptor-gamma coactivator-1α (PGC-1α) gene. Faradilla and Rahman (2020) studied the effect of garlic extract on the histopathological features of the pancreas and found that it alters the activity of the cells in the islets of Langerhans. S-allylcysteine sulfoxide, found in garlic extract, has a stimulatory effect on the insulin secretion ability of pancreas cells (Liu et al., 2005; Islam and Choi, 2008). Bhattacharya et al. (2019) reported that S-allylcysteine sulfoxide has a stimulatory effect on both insulin secretion and glucose transporter-4. Ojo et al. (2021) mentioned the negative consequences of protein tyrosine phosphatase 1B (PTP1B) in insulin signaling pathways. They suggested that the inhibitory effect of garlic extract on PTP1B may help in type 2 diabetes treatment.

4.3.2 Effects of Garlic on Insulin Sensitization

Blahova et al. (2021) noted that the hypoglycemic effect of garlic is due to the prevention of insulin inactivation by the sulfhydryl group and increased insulin sensitization. Padiya et al. (2011) found that garlic improves insulin sensitivity in fructose-fed rates. The use of garlic-containing herbal medicine increases insulin secretion and sensitivity while decreasing insulin resistance (Parham et al., 2020). According to Zhai et al. (2018), alliin present in intact garlic is responsible for this effect. Garlic oil or diallyl trisulfide also enhances the glucose tolerance as well as increases the insulin sensitivity (Liu et al., 2005). Allixin, isolated from dried garlic bulbs, also helps normalize hyperglycemia in mice by improving insulin resistance without any changes in plasma adiponectin levels (Kirkman et al., 2006).

4.3.3 Complications of Diabetes Mellitus and Garlic Use

Higher glucose levels in diabetes mellitus patients caused temporary as well as long-term effects on various organs. It may cause heart disease, stroke, renal and ocular problems and microvascular abnormalities, such as neuropathy, foot problems, cutaneous problems, gastrointestinal problems and mental illness. Kirkman et al. (2006) and Stettler et al. (2006) noted that various micro- and macrovascular diseases are linked to type 1 and type 2 diabetes. Lindstedt et al. (2021) reported that the consumption of aged garlic extract in cases of arterial arteriolosclerosis improved endothelial cell function and blood perfusion to peripheral tissue. It acts directly on the arterial wall (Ribeiro et al., 2021) and increase cystathionine-ϒ-lyase expression in the myocardium (Tsai et al., 2015). Garlic oil use could lighten diabetic liver damage, as evidence by the regulation of liver peroxisome PGC-1α and irisin-encoding gene experession (Eser et al., 2021; Dragos et al., 2020; Yuvashree et al., 2020). Diabetic rats treated with garlic showed lower urinary protein levels and reduced expression of renal CD36, podacalyxin and NGAL compared to the control group, indicating nephroprotection in garlic-treated diabetic rats (Dragos et al., 2020; Yuvashree et al., 2020). In a clinical trial on gestational diabetes mellitus, there was a significant increase in insulin resistance. It was concluded that probiotic bacteria improved the antitoxidant effect of garlic in gestational diabetes milletus patients by converting the glycopyranoside in fresh garlic to glucofuranoside (Si L. et al., 2019).

Garlic's active components affect the Nuclear Factor kappa B (NF-kB) pathway, leading to downregulation of its mRNA expression. These effects reduce retinal oxidative stress and diabetic retinopathy (Al-Brakati, 2016; Lotfi et al., 2021). In an extensive study, Kommula et al. (2020) found that the use of 3% garlic in the daily diet of prediabetic Wistar rats could prevent functional and structural abnormalities of the retina. Vascular endothelial growth factor and glial fibrillary acid protein expression initiate the processes of angiogenesis and glial activation in the diabetic retina. Increased expression of carboxymethyl lysine (CML-KLH) and 4-hydroxynanoenol (4-HNE) indicates glycation and retinal oxidative stress (Kommula et al., 2020). The antioxidant and anti-inflammatory effects of aqueous garlic extract on the retinal tissues of diabetic rats have been reported by Ziamajidi et al. (2022) and Lotfi et al. (2021). They found that garlic extracts decrease the retinal oxidative stress. The expression levels of TGF-β2 and IL-1β were significantly reduced in diabetic rats supplemented with garlic. Recently, Afarid et al. (2022) noted that garlic intake could significantly improve visual acuity, decrease central macular thickness and reduce intraocular pressure in a randomized clinical trial involving 117 eyes of diabetic patients, clearly indicating that garlic can be used as a complementary treatment for diabetic macular edema (Afarid et al., 2022).

4.4 ANTICANCER POTENTIAL OF GARLIC

In a healthy body, the division, differentiation and death of cells are carefully regulated. Cancers start as a single cell that has lost control of normal growth and replication processes. Transformation of a normal cell to a cancerous cell involves distinct phases, i.e. initiation, promotion, progression and further evolution to malignant tumors. These steps can be modified by dietary habits. Dietary habits account for one-third of cancer deaths (Boivin et al., 2009). Individuals eating more fruits and vegetables daily have about half the risk of developing various cancer types, mainly those of gastrointestinal tract (WCRF/AICR, 2007). Fruits and vegetables contain various natural chemical compounds, which are potent modifiers of chemical carcinogenesis. One of the first plants reported to possess antitumor compounds was garlic. Garlic is one of the most ancient spice plants reputed to have an effect on cancer. As recorded around 1550 BC in the Ebers Papyrus, garlic was applied externally for the treatment of tumors by ancient Egyptians and internally by Hippocrates and Indian physicians (Hartwell, 1967, 1968). Cancer chemoprevention with garlic and its different constituents has been the subject of various recent scientific studies (Reuter et al., 1996; Sengupta et al., 2004; Khanum et al., 2004; Ariga and Seki, 2006; Milner, 2006; El-Bayoumy et al., 2006; Stan et al., 2008; Herman-Antosiewicz and Singh, 2004; Herman-Antosiewicz et al., 2007; Siess et al., 2007; Ngo et al., 2007; Shukla and Kalra, 2007; Stan et al., 2008; Powolny and Singh, 2008; Nagini, 2008; Yung-Lin et al., 2013; Jikihara et al., 2015; Jiang et al., 2017; Li et al., 2018a; Shang et al., 2019; Zhou et al., 2020; Zhang et al., 2020; De Greef et al., 2021; Pandey et al., 2023).

Interestingly, China provides an ideal "Field Laboratory" for epidemiological studies of cancer incidence. Stomach cancer was found to rank higher for males and females in cancer mortality (Wang et al., 1985; Lau et al., 1990) than other cancer incidence in China (Mei et al., 1982). They suggested that consumption of garlic may have inhibited nitrate reduction by bacteria. Subsequently, the lower gastric nitrite (a nitrosamine precursor) concentration may reduce the risk of developing stomach cancer. You et al. (1989) identified that cigarette smoking, use of salty foods and moldy grains are associated with an increased risk of stomach cancer. A significant reduction of stomach cancer risk was found to be associated with an increased consumption of garlic, scallicens and Chinese chives (You et al., 1988). Galeone et al. (2006) investigated the role of allium's vegetables in the etiology of various neoplasms. They analyzed the relationship between the frequency of onion and garlic use and cancer at several sites in Italian and Swiss case-control studies. They reported an inverse relationship between the frequency of use of garlic and the risk of several common cancers. Abdullah et al. (1988) observed that natural killer cells from the raw garlic and aged garlic groups destroyed 139% and 159% more tumor cells, respectively, than those from the control group. Aged garlic extract is a good chemopreventive agent for colorectal cancer. Aged garlic extract could prevent tumour formation by inhibiting angiogenesis through the suppression of endothelial cell mortality, proliferation and tube formation (Matsuura et al., 2006).

Aflatoxins are naturally occurring toxic metabolites produced by *Aspergillus flavus* and related fungi. Among the 16 or more toxic compounds secreted by these fungi, aflatoxin B1 (AFB1) is most potent in causing mutations in various microorganisms as well as mammalian and human cells. It is both toxic and hepatocarcinogenic to a variety of animals and humans in Asia and Africa (Stoloof, 1976; Wong and Hsieh, 1976; Perkin et al., 1988). Tadi et al. (1991a, 1991b) reported that two organosulfur compounds of garlic, ajoene and diallyl sulfide, affected AFB1 metabolism and DNA binding by inhibiting phase I enzymes and, therefore, act as potential cancer chemopreventive agents.

Garlic is one of the best natural sources of germanium. It is of interest to note that this trace metal has also been reported to prevent and cure cancer. Garlic is also an excellent source of selenium, which has potential therapeutic value in cancer treatment (Bolton et al., 1982; Lawson, 1993). Se-methylselenocysteine is the major seleno-compound in the garlic bulb. Epidemiological studies have indicated a negative relationship between selenium intake and the incidence of certain cancers. Blood or plasma levels of selenium are usually lower in patients with cancer than those without the disorder. In a mammary tumour model, se-methylselenocysteine was shown to be the most effective seleno-compound so far in reducing tumours (Whanger, 2004).

Although garlic has been shown to be chemopreventive against cancer, its mode of action is not fully understood. Several mechanisms have been suggested. A potential carcinogen may exert its effects by producing nuclear damage. Wargovich and Goldberg (1985) studied the effect of a sulfide and disulfide component of garlic on several chemical carcinogens. DMH (1,2-dimethyhydrazine) and NMBA (N-nitroso methyl benzylamine) are procarcinogens, which are metabolized in the liver to carcinogenic metabolites, an effect mediated by their binding to the nucleus. Pre-administration of diallyl sulfide prevented nuclear damage induced by DMH and NMBA. These observations suggest that diallyl sulfide might inhibit the conversion of procarcinogens to ultimate carcinogens, an important factor of the initiation phase of carcinogenesis. Chung et al. (2002) observed that garlic juice reduced the endogenous formation of the carcinogen N-nitrosodimethylamine (NDMA) in young humans. Recently, Singh et al. (2006) have reported that consumption of garlic powder induced a 35–60% reduction in DNA damage induced by NDMA in rats.

Glutathione-s-transferase (GST) is an important enzyme that conjugates electrophils and assists in the detoxification of many carcinogens. Compounds that increase GST activity have been found to inhibit carcinogen-induced neoplasia (Wattenberg, 1983; Sparnins et al., 1988). Oral administration of methyl trisulfide and diallyl disulfide stimulated glutathione-s-transferase activity, which increased with time and showed a time lag of 18 hours or more (Sparnins et al., 1986; Sumiyoshi and Wargovich, 1989). These studies suggest that the stimulation of glutathione-s-transferase activity in the liver and other target organs of carcinogens by garlic-derived organosulfur compounds may be responsible for their protective effects in chemical carcinogenesis and on different stages of carcinogenesis.

Oxygen radicals may play a role in tumor promotion. Various antioxidants, trace radicals, scavengers, stimulation of super oxide dismutase, glutathine peroxidase inhibitors of cyclooxigenase and lipooxygenase enzymes were found to modulate the oxidative stress and the formation of molecular species involved in tumor promotion. Perchellet et al. (1986) demonstrated garlic oil stimulated glutathione peroxidase activity and inhibited the decrease in the intracellular ratio of reduced to oxidize glutathione produced by TPA in epidermal cells. Garlic oil was also found to inhibit lipooxigenase enzymes, which are involved in TPA-stimulated arachidonic acid metabolism (Block et al., 1988). Li et al. (1995) investigated the effect of aged garlic extract (AGE) and two of its components, s-allylcysteine (SAC) and s-allyl mercaptocysteine (SAMC), on human breast cancer cells MCF-7 and MCF-7-s. They observed an antiproliferative response to SAC and SAMC and an alteration in glutathione levels without significant concurrent changes in the glutathione-metabolizing enzymes. Xiao et al. (2003) found that SAMC exerts antiproliferative effects by binding directly to tubulin and disrupting the microtubule assembly, thus arresting cells in mitosis and triggering JNK1 and caspase-3 signaling pathways that lead to apoptosis.

Xu et al. (2004) reported a novel aspect into the understanding of the molecular mechanism of potential antitumor action of ajoene. Ajoene-treated leukemia cell line HL60 cells were arrested in the G(2)/M phase, and the total amount of cytosolic proteasome increased. Proteasome plays a key role in the regulation of many cellular processes. Oommen et al. (2004) found that allicin inhibited the growth of cancer cells of murine and human origin. Allicin induced the formation of apoptotic bodies, nuclear condensation and a typical DNA ladder in cancer cells. They have observed that allicin activated the caspases-3, -8 and -9, as well as the cleavage of poly (ADP-ribose) polymerase. Xiao et al. (2006) observed that diallyl trisulfide, found in garlic bulbs, induces apoptosis in PC-3 human prostate cancer cells inducing the proteins Bax and Bak. Chu et al. (2006) studied the effect of the garlic derivatives S-allylcysteine and S-allylmercaptocysteine on the inhibition of cancer cell invasion through the restoration of E-cadherin expression. They suggested that these compounds might be potential therapeutic agents for suppressing androgen-independent prostate cancer.

It seems that garlic-derived organosulfur compounds could have effects on different stages of carcinogenesis. It is now recognized that a wide range of human cancers, perhaps as many as 80–90% is attributed to environmental factors (Doll, 1977; Doll and Peto, 1987). Garlic has been shown to inhibit the growth of transplantable tumors and to reduce the incidence of certain spontaneously occurring tumors. Components of garlic also inhibit the activity of diverse chemical carcinogens during both the initiation and promotion phases of carcinogenesis (Lau et al., 1990; Tadi et al., 1990; 1991a, 1991b; Milner, 1996).

Kodera et al. (1989) isolated a new allixin (phytoalexin) synthesized by garlic. Being a phenolic compound, it is an effective inhibitor of phospholipid metabolism stimulated *in vitro* by the tumor promoter. Yamasaki et al. (1991) observed that allixin plays an important role in cancer prevention. Lau et al. (1991) and Patya et al. (2004) studied the effect of organosulfur compounds of garlic on the growth of animal tumors. According to Patya et al. (2004), allicin exhibits immune-stimulatory and antitumor properties. The immune-stimulatory effect of allicin is mediated by redox-sensitive signaling, such as the activation of p2(ras). The antitumor effect of allicin is related to its immune-stimulatory properties. Allicin inhibits the apoptosis of macrophages in depleted nutritional states through the mitogen-activated protein kinase/extracellular signal-regulated kinase pathway (Cho et al., 2006). Druesne et al. (2006) observed that diallyl disulfide (DADS), found in garlic bulbs, inhibits colon tumor cell proliferation. They suggested that 200 μM of DADS increases histone acetylation, CDKN1A mRNA and p21 (waf1) protein levels and induces G2/M cells cycle arrest in tumor cells. Agarwal et al. (2007) observed that pretreatment of rats with 50–100 mg garlic oil/kg body weight for a week significantly protects free radical injury and kidney cancer induced by ferric nitrilotriacetate (Fe-NTA). This is due to the inhibition of hepatic tumour markers, viz., ornithine decarboxylase activity and DNA synthesis. It is therefore obvious that the organosulfur compounds of garlic extend their protective anticancer effects in three ways: (i) a direct inhibition of tumor cell metabolism, (ii) inhibition of the initiation and/or promotion phases of carcinogenesis and (iii) modulation of the host response by augmenting macrophages and T-lymphocytes function.

Boivin et al. (2009) studied the inhibitory effects of 34 commonly consumed vegetables on the proliferation of eight different tumor cell lines obtained from stomach, kidney, prostate, breast, brain, pancreatic and lung cancers. The extracts from cruciferous vegetables, particularly those from the genus *Allium*, inhibited the proliferation of all tested cancer cell lines. There was a negligible effect on the growth of normal cells. Among all the vegetables tested, the extract from garlic was the strongest inhibitor of tumour cell proliferation, with complete growth inhibition of all tested cell lines. An extensive report by the World Cancer Research Fund concludes that there is convincing evidence of "decreased risk" of stomach cancer associated with the consumption of allium vegetables and of colorectal cancer associated with the consumption of garlic (WCRF/AICR, 2007). The European Prospective Investigation into Cancer and Nutrition (EPIC) study, carried out in 23 centers across ten European countries and including 519,978 subjects (aged 35 to 70 years), revealed that the consumption of onions and garlic probably reduces the risk of intestinal and stomach cancers but is probably not associated with cancers of the lung, prostrate or breast (Gonzalez and Riboli, 2006). Galeone et al. (2006, 2008) also concluded the same findings in smaller European studies. Millen et al. (2007), in his study, reported that deep-yellow vegetables, dark-green vegetables, onions and garlic are associated with a reduced risk of colorectal adenoma.

In a clinical trail, 0.4% ajoene cream for the treatment of basal cell carcinoma was applied in 21 patients with a total of 25 lesions. The size of each lesion was measured once a month (Tilli et al., 2003; Hassan, 2004). After six months, the basal cell carcinomas were completely removed by surgery, and the activity of the tumor cells examined from the earlier treatment 21 showed an 84% reduction in size within six months. The tumor size reduction ability of ajoene is associated with the decrease in the number of cells expressing B-cell lymphoma Z (BCl-2) (Tilli et al., 2003). Zhang et al. (2008), in a clinical trial in China, reported that progressive gastric adenocarinoma was detected in 80 patients. They were divided into two groups, with 40 patients in each group. Forty-eight hours before the operation, allicin was infused via gastroscopy into the lesion region of patients in the allicin group, while normal saline was infused in the control group. It was noted that local application of allicin through gastroscopy caused significant inhibition of cell growth and proliferation in progressive gastric carcinoma.

The potent antibiotic and cytotoxic effects of allicin have thus limited to topical or *in vitro* applications since allicin reacts rapidly with free thiol groups (glutathione, abundant in plasma and in cells) and thus disappears from the circulation shortly after injection. Arditti et al. (2005) have given a new technique to overcome this limitation in antitumor therapy, which involves an *in situ* generation of allicin on the targeted cells. The alliinase enzyme is first targeted to a tumor by conjugating the enzyme to a carrier, a monoclonal antibody specific to a tumor-associated surface antigen. Alliin is then introduced into the circulation. In this way, cytotoxic allicin is formed only at the site of alliinase localization on the tumor cell surface. This type of drug delivery is based on the concept of antibody-directed enzyme prodrug therapy (Miron et al., 2003). Arditty et al. (2005) used this therapy to kill human lymphocytic leukemia tumor cells *in vivo*. Prager-Khoutorsky et al. (2007) showed that allicin is a potent, microtubule-disrupting reagent that interferes with tubulin polymerization by reacting with tubulin SH groups. In this way, allicin is able to inhibit cell polarization, migration and division. A similar drug delivery system was used by Zhang et al. (2007). They used polybutylcyanoacrylate (PBCA) nanoparticles of diallyl trisulfide, which has antitumor activity in hepatocellular carcinoma in mice. The diallyl trisulfide polybutylcyanoacrylate nanoparticles, measuring 113.4 nm, were prepared by emulsion polymerization with PBC as the carrier and slowly released the diallyl trisulfide *in vivo* (Zhang et al., 2007).

Ajoene and the garlic oil component diallyl trisulfide are used as an experimental antileukemia drugs (Hassan, 2004; Hodge et al., 2008). These compounds alter the carcinogen metabolism and suppress the growth of cancer cells by inducing cell cycle arrest, promoting apoptosis and suppressing of angiogensis and metastasis (Dirsch et al., 2002). Diallyl polysulfides containing three to five sulfur atoms are more potent than diallyl disulfide in inhibiting cancer cell proliferation (Nishida et al., 2008). Chemical substances in an organism are detoxified mainly by two enzyme systems known as phase I and phase II and then cleared from the body. Phase I enzymes oxidize the chemical substances through the cytochrome P450 enzyme system, producing cytotoxic reactive intermediates. On the other hand, phase II enzymes eliminate the toxic chemical substances by conjugating the intermediates generated by phase I enzymes to glutathione or glucuronic acid. Diallyl sulfide/polysulfides caused a significant increase in the phase II enzymes glutathione S-transferase and quinine reductase in rats (Powolny and Singh, 2008).

Iciek et al. (2012) reported the antitumor properties of organosulfur compounds, including allicin, diallyl disulfide, alliin, diallyl sulfide, allyl mercapton and S-allyl cysteine, isolated from garlic. In addition, garlic powder inhibited the DNA damage caused by N-nitrosodimethylamine in the liver by 35% when administered to rats, an effect attributed to the high concentration of alliin—up to 60%—in the samples (Singh et al., 2006). Yung-Lin et al. (2013) discussed how natural food factors regulate cell survival or death by autophagy in cancer cells. Garlic not only induces apoptosis but also triggers autophagy in cancer cells. Autophagy, also known as type II programmed cell death, provides a new strategy in cancer therapy. Jikihara et al. (2015) noted the anticancer potential of alcoholic extracts of *A. sativum*. They found cytotoxic activity of *A. sativum* aqueous preperations in rodents with 1,2-dimethylhydrazine (DMH)-induced carcinogensis and human colorectal tumor cell proliferation. Apoptosis was not induced by the alcoholic garlic extract. Instead, it slowed down cell cycle progression by decreasing the expression of cyclin B1 and Cdk1 in human colorectal cancer cells. The effects of garlic extract on the growth of human breast cancer cell lines (MCF-7), prostate (PC-3), liver (Hep-G2) and colon (Caco-2) cells, as well as mouse macrophage cells, were investigated by Bagul et al. (2015). The exposure of Hep-G2, MCF-7, TIB-71 and PC-3 cells with crude garlic extract decreased cell proliferation by 80–90% but only 40–55% in Caco-2 cells. It caused growth arrest and a fourfold increase in caspase activation (apoptosis) in PC-3 cells

(Bagul et al., 2015). The *in vivo* and *in vitro* effects of diallyl trisulfide were studied against gastric cancer cells SGC-7901. Diallyl trisulfide inhibited SGC-7901 cell growth by inducing apoptosis and amplifying the MAPK route by phosphorylating JNK, ERK and P38. Diallyl trisulfide therapy increased cytokine release, such as TNF and IFN, which promoted the inflammatory system of the host in cancer care (Jang et al., 2017).

Allicin, an active component of crushed garlic clove, is effective against stomach cancer. Allicin functions by halting the cell cycle in the G2/M phase and apoptosis in gastric tumor cells, while normal stomach cells remain unaffected. TGF-alpha (TGF-α) and its epidermal growth factor receptor (EGFR) are inhibited by allicin, which causes cyclin E and cylin D1 to be downregulated, and the cells in the G2 phase were not able to enter the M phase. Allicin also caused the inhibition of DNA damage by reactive oxygen species (ROS). It phosphorylates P53 and P21 proteins. P21 then inhibits the cyclin-dependent kinase 4/6 (CDK 4/6)-cyclin D complex, preventing the cell from entering the G2/M phase. The P21 polypeptide can block the P21 CDK2 and P21-proliferating cell nuclear antigen (PCNA) assemblages, causing the CDK1-cyclin B1 complex to diminish and G2/M cell stage to be arrested. The elevation of the proapoptotic polypeptide BAX, associated with B-cell lymphoma 2 (BCL2), induced by allicin, promotes cell death through the mitochondrial route.

A greater BAX/BCL2 ratio causes cytochrome C to be released into the cytoplasm, accompanied by a caspase-9 involvement. First, caspase-3 is triggered, which results in poly ADP-robose polymerase (PARP) and apoptosis-inducing polymerase (AIP) activation. The intrinsic stimulation of p38 mitogen-activated protein kinase (p38 MAPK) activity was greater in allicin-induced gastric tumor cells. The synthesis of the Fas membrane-spanning peptide and P53 may be increased if p38 MAPK is increased. The Fas protein is part of the tumor necrosis factor, or TNF receptor, superfamily and contains three domains (extracellular, cytoplasmic and transmembrane). The association of the Fas cysteine-rich signal peptide with the Fas ligand (FasL) is a critical regulator of tumor growth (Luo et al., 2016). Stimulation of caspase-8 triggers caspase-3 and caspase-7 in the mitochondrial-independent route to trigger apoptosis. Caspase-8 is responsible for the liberation of Cyt c from mitochondria in the mitochondrial-dependent route (Burz et al., 2009). Allicin can also result in cell death through a process devoid of caspases. Mitochondria generate apoptosis effectors, such as apoptosis-inducing factor (AIF) and endonuclease G (Endo G), and AIF is then teleported to the nuclei, causing cell death through breaks in the nuclear material strands. Allicin can cause apoptotic cell death by boosting unbound Ca^{2+} cation concentrations and inducing endoplasmic reticulum (ER) strain (Luo et al., 2016).

Allicin isolated from garlic cloves was reported to suppress colorectal cancer metastasis by enhancing immune function and preventing the formation of blood vessels, as well as inhibiting survival gene expression to promote cancer cell apoptosis. It also can enhance the treatment of pancreatic cancer by reversing gene silencing and restraining cancer cell proliferation (Chhabria et al., 2015). Zhang et al. (2015) revealed that allicin can prevent the profileration of gastrointestinal cancer cells (MGC 803) and induce apoptosis, which can be done by enhancing p38 expression and cleaved caspase-3. Gruhlke et al. (2016) noted that tumor cells can be killed when incubated in an allicin solution. Li et al. (2018b) found that raw garlic was the most effective and highly specific anticancer drug when compared to 33 raw vegetable extracts against different cancer cells without affecting the noncancerous cells. Shang et al. (2019) reported that the anticancer potential of garlic extracts was attributed to the inhibition of cell growth and proliferation, regulation of carcinogen metabolism, stimulation of apoptosis, prevention of angiogenesis, invasion and migration, thus reducing the anticancer agent's negative effects.

In China, lung cancer has the highest incidence and mortality rates (Torre et al., 2015). In the past 30 years, lung cancer mortality in China increased by 464.84% (She et al., 2013). Smoking, exposure to secondhand smoke and air pollution are the major risk factors of lung cancer (She et al., 2013). Myneni et al. (2016) noted that the reduced risk of lung cancer associated with low/no exposure to indoor air pollution and the consumption of raw garlic might be greater than the risk reduction associated with any one of these factors alone. They found an additive interaction between supplement use and raw garlic consumption in lung cancer. According to them, anticancer properties are mainly attributed to bioactive organosulfur compounds. Garlic contains other nutrients, including flavonoids, oligosaccharides, arginine, potassium, vitamin C and selenium, that may contribute to its overall beneficial effects (Myneni et al., 2016).

Li et al. (2018a, 2018b) reported that raw garlic extract injection had a novel therapeutic effect against even the most aggressive models of malignancy in mice. Ingestion of the same amount of raw garlic extracts failed to offer any meaningful effect on the same model of malignancy. This activity was attributable to multiple garlic compounds that were partially heat-sensitive, organic-soluble and smaller than 3,000 Daltons. The treated mice did not display any adverse side effects in terms of behavior or physical appearance from receiving daily injections of raw garlic extract. The extremely potent anticancer activity of raw garlic extract was only achieved through direct exposure to cancer cells. They noted that raw garlic extract injections killed only the cancer cells, not the normal cells. According to them, this selectivity is possibly due to the metabolic properties of cancer cells that make them targets for garlic compounds (Li et al., 2018a, 2018b).

The anticancer potential of garlic extract is mediated by increased endoplasmic reticulum stress (Petrovic et al., 2018). They examine the molecular and cellular activities of a simple homemade ethanol-based garlic extract. Garlic extract inhibits the growth of several different cancer cells *in vitro*, as well as cancer growth *in vivo* in a syngeneic orthotopic breast cancer model. The garlic extract was fractionated using solid-phase extraction. Anticancer potential was lost during freeze or vacuum drying, suggesting that the main anticancer compounds in garlic extracts are volatile. The anticancer activity was stable for more than six months at –20°C. Garlic extract enhanced the activities of chemotherapeutics, as well as MAPK and P13K inhibitors, and affected hundreds of cellular signaling proteins, including changes in vital cell signaling cascades regulating proliferation, apoptosis and the cellular redox balance. Reduced proliferation of cancer cells treated with garlic extract is at least partly mediated by increased endoplasmic reticulum stress (Petrovic et al., 2018).

In a meta-analysis, Zhou et al. (2020) found that garlic intake can lower the colorectal cancer. According to the World Cancer Research Fund/American Institute for Cancer Research, garlic consumption has a potentially favorable effect against colorectal cancer (Wiseman et al., 2007). S-allylmercapto-cysteine, the water soluble derivative of garlic, has been found to display antiproliferative capacity in numerous cancer cell lines, and it has also shown tumor-inhibiting effects in *in vivo* conditions (Thomson and Ali, 2003; Liang et al., 2011). Garlic can enhance the anticancer ability by activating the antioxidant transcription expression of Nrf2 and the downstream gene NQ01 (Li et al., 2017). Aged black garlic can exert its anticancer function by suppressing the proliferation of cells (Jikihara et al., 2015).

Zhang et al. (2020) and Pandey et al. (2023) extensively discussed the anticancer potential of garlic preperations. The most common cancer in women worldwide is breast cancer. More than 2 million new cases of female breast cancer were anticipated, along with 684,996 cancer-related fatalities worldwide in 2020. Today, proliferation of breast cancer has been reduced by using different available therapeutics. However, these therapies are still associated with life-threatening toxicities. Garlic is now recognized to have potent anticancer efficacy. Several organic and sulfur components present in garlic cloves affect multiple stages of cancer cell proliferation, development, growth, migration, invasion and metastasis by interfereing with cell cycle regulation, inhibiting cell signaling pathways, inducing apoptosis, promoting autophagy and enhancing antioxidant potential (Shang et al., 2019; Zhang et al., 2020; De Greef et al., 2021). The therapeutic potential of garlic nanoformulations was higher than that of the same component alone. Allicin nanoparticles showed stronger anti-angiogensis impact than allicin by itself (Shafeeque and Hashim, 2018). The use of the nanoformulation drug delivery system can increase the bioavailability of medications that are poorly water-soluble, allow for the co-delivery of several medications, prevent drug toxicity in normal cells and extend the duration of therapeutic activity. Reactive oxygen species and cytotoxicity investigations revealed an increased antitumor potential of diallyl disulfide when incorporated into solid lipid nanoparticles, compared to the free diallyl disulfide drug in breast cancer cells (Nkrumah-Elie et al., 2012). The growth of MCF-7 human cancer cells was inhibited by an aqueous preparation of allicin and garlic powder. The growth-inhibitory impact of allicin was closely related to the reduction in intracellular glutathione (GSH) levels in both MCF-7 (luminal-A) and HCC-70 (triple negative) breast cancer cells (Hirsch et al., 2000; Chen et al., 2016a; Rosas-Gonzalez et al., 2020). Alliin and allicin reduced the cell viability of MCF-7 (luminal A) and HCC-70 (triple negative) breast cancer cells. Allicin had more pronounced cytotoxicity and anti-proliferative activity against both cell lines by lowering mitochondrial membrane potential, inducing apoptosis and enhancing the expression levels of various caspases. It also increased the levels of proapoptotic proteins (p21 Noxa and Bak) while downregulating the expression of the antiapoptotic protein Bcl-xL (Tsubura et al., 2011; Milczarek, 2020; Fatima et al., 2021; Catanzaro et al., 2022).

Diallyl disulfide showed cytotoxicity towards the MCF-7 (ER-positive human breast cancer) cells, T47D (differential epithelial sub strain), KPL-1 (created from the malignant effusion of a breast cancer patient), MDA-MB-231 and MKL-F (ER-negative) human breast cancer cells (Na et al., 2012; Hahm et al., 2021). Cell cycle arrest (Sub-G1 phase) and apoptotic induction was noted in MDA-MB-231 cells, which was accompanied by modulations in the expression levels of numerous proteins, such as Bax, BcL-xL and caspase-3. Diallyl trisulfide treatment significantly increased growth arrest at the G2/M phase in both MCF-7 and MDA-MB-231 cells. In addition, diallyl trisulfide administration dramatically elevated the ser10 phosphorylation of histone H3 (mitotic marker) in both breast cancer cells. Diallyl trisulfide did not affect SLIT/R0B0 signaling in breast cancer cells (Huang et al., 2015). Diallyl trisulfide–treated MCF-7 cells demonstrated enhanced AP-1 DNA binding activity, which was suppressed by NAC and the JNK inhibitor. The cells treated with C-jun siRNA were resistant to proteolytic poly ADP-ribose polymerase cleavage in diallyl trisulfide–treated cancer cells. Female Balb/c mice given 5 mol/Kg diallyl trisulfide orally were unable to generate human MCF-7 cell tumor xenografts (Nkrumoh-Eli et al., 2012). Xiao et al. (2014) also reported diallyl disulfide cytotoxicity towards MCF-7 cells with reduced cell proliferation and apoptosis induction via intrinsic signaling pathways associated with upregulated protein (Bax, Bad and caspases) expression levels and downregulated BCl-2 protein expression levels.

Diallyl disulfide treatment resulted in reduced cell growth of several human triple-negative breast cancer cell lines (MDA-MB-231, BT-549 and MDA-MB-468) by regulating the β-catenin signaling pathway. Increased expression of Bax, caspase-3 and caspase-9, as well as decreased levels of Bcl-2, matrix metalloproteinase-9 and β-catenin siRNA, were associated with the induction of apoptosis and inhibition of cell proliferation (Malki et al., 2009; Nkrumah-Eli et al., 2012; Li et al., 2018b). In MCF-10A cells (typical human mammary epithelial cells), diallyl trisulfide reduced the ability of benzo[a]pyrene to cause cancer by preventing cell division, causing cell cycle arrest at the G2/M phase and inhibiting the production of ROS and DNA damage (Kiesel and Stan, 2017). Liu et al. (2017) reported that diallyl trisulfide treatment induced cell cycle arrest (G2/M phase) and apoptosis, which drastically reduced MCF-7 cell viability and not affected the MCF-12A normal human breast epithelial cells. Enhanced translocation of p53 from the cytoplasm into the nucleus was associated with the activation of apoptosis, along with the upregulation of the Bax and p53 protein expression levels. Akt and Bcl-2 protein expression levels were further inhibited, whereas FAS and cyclin D1 protein expression levels were elevated. Trx-1 expression and Trx-1 reductase enzyme activity were decreased by diallyl trisulfide in breast cancer cells (MDA-MB-231, ZR-75–1, Hs 578T and MCF-7) (Malla et al., 2021). The cysteine residues in thioredoxin-1 (redox-active disulfide/dithiol) containing 12 KD proteins are converted into an oxidized disulfide bond (intramolecular Trx-1-S2) configuration through a chemical process. Trx-1 overexpression in tumor cells speed up the growth of tumors and reduce patient survival. Trx1 overexpression in tumor cells caused the JNK/p38 signaling pathway-associated apoptosis signal-regulating ki nase 1 (ASK1) expression to be suppressed, which in turn suppressed apoptosis.

In vivo studies on diallyl trisulfide–treated breast cancer xenografts demonstrated reduced STAT3 expression. Diallyl trisulfide treatment decreased the incidences of breast cancers in xenografted MDA-MB-231 tumor models via reducing MMP2 and MMP9 expression levels (Wei et al., 2017; Liu et al., 2018b; Vijayakumar et al., 2019). In a rat model of N-methyl-N-nitroso-urea-induced mammary tumors, the use of garlic powder inhibited the growth of tumors and proved the fact that diallyl disulfide is more active than allyl monosulfide and diallyl sulfide (Schaffer et al., 1996).

Clinical trials of garlic products and constituents were studied in different groups. Five grams of crushed garlic (raw) enhanced the expression of several genes, including the proto-oncogene c-Jun, the aryl hydrocarbon receptor nuclear translocator, HIF-1A (hypoxia-inducible factor 1A), oncostatin M and the viral oncogene homolog (V-rel avian reticuloendothelosis) in a clinical trial involving 17 volunteers. Garlic components are multifaceted and include the activation of genes associated with immunity, xenobiotic metabolism and apoptosis in humans and monocytes (Charron et al., 2015). Earlier, similar findings were done in clinical research that involved 20 cancer patients and noted that raw garlic and allicin had cytotoxic effects against a variety of cancer types (Bespalov et al., 2004).

In stomach cancer patients, a treatment of 200 mg aged garlic extract and 1 mg of steam-distilled garlic oil was administered twice daily for a period of seven years. In this study, 3,365 participants were involved to investigate the incidence and mortality of stomach cancer using a 22 factorial design. Follow-up period was initially 7.3 years and was later extended to 22.3 years (Li et al., 2019; Guo et al., 2020). It was noted that garlic had a more protective impact on gastric cancer prevention when combined with abstinence from alcohol.

4.5 ANTIOXIDANT ACTIVITY OF GARLIC

Reactive oxygen species (ROS), including hydrogen peroxide, superoxide radical, hydroxyl radical and peroxynitrite, are responsible for cellular damage to nucleic acids, proteins and lipids, which causes cancer, chronic inflammation and cardiovascular diseases. Oxidative stress can lead to endothelial remodeling, tissue damage and, eventually, atherosclerosis (Canugovi et al., 2019). Augusti (1996) found that aqueous garlic extracts protect tissue against oxidative damage caused by nicotine. Later, Fukao et al. (2008) reported that diallyl trisulfide helps heal liver damage caused by carbon tetrachloride. Garlic has also been shown to protect against lipid-related oxidation through the actions of S-ethyl cysteine and N-acetyl cysteine (Tsai et al., 2005). Prasad et al. (1995) noted that hydroxyl redicals are scavenged by allicin. Garlic oil enhances glutathione peroxidase activity and prevents the reduction in the intracellular ratio of reduced to oxidized glutathione caused by 12-O-tetradecanoylphorbol-13-acetate (TPA) in epidermal cells (Perchellet et al., 1986). Garlic oil was also discovered to block lipoxygenase, an enzyme essential for TPA-stimulated arachidonic acid metabolism (Belman, 1985).

Allium sativum possesses high antioxidant properties. Raw, uncooked garlic has an intense antioxidant potential (Locatelli et al., 2017). Jang et al. (2018) noted that both aged garlic and nonaged garlic extracts of water, ethyl alcohol or dichloromethane showed distinct antioxidant effects as measured by different radical scavenging tests. Multiclove garlic isolates and single-clove garlic both exhibited a substantial increase in scavenging activity in the DPPH test and against superoxide redicals. Single-clove garlic was more resistant to carbon tetrachloride-induced hepatotoxicity than multiclove garlic isolates. It may be an effective substitute drug for severe oxidative hepatoxicity (Naji et al., 2017).

Borek (2001) found that aged garlic extract and its various constituents have proven an array of antioxidant and radio-protective effects. The use of different aged garlic products protect white blood cells from radiation damage, liver cells from lipid peroxidation and vascular endothelial cells from oxidant injury and enhance antioxidant enxyme systems in cells. Ebid et al. (2006) found that reactive oxygen species seem to be at the core of many ailments. The antioxidant effect of garlic occurs through the modulation of reactive oxygen species and by increasing glutathione and cellular antioxidant enzymes. Asdaq and Inamdar (2011b) reported that regular garlic intake promotes internal antioxidant activities and reduces oxidative adverse effects either by increasing the endogenous antioxidant synthesis or by reducing oxygen-free radical species. Garlic protects aginst gentamycin- and acetaminophen-induced hepatotoxicity by improving antioxidant status and regulating oxidative stress (Wallock-Richards et al., 2014). Aged garlic extracts rich in flavonoid phenol and different sulfur compounds show high radical scavenging activity (Jang et al., 2017). Aged garlic extracts act by stimulating the expression of different antioxidant enzymes, such as glutamate-cysteine ligase modifier (GCLM) and heme oxygenase-1(HO-1), through the nuclear factor erythroid 2-related factor 2 (Nrfr2) antioxidant response element pathway, which is responsible for protecting human endothelial cells against oxidative stress (Liu et al., 2018b). Chen et al. (2016b) reported that alliin, the major sulfur compound isolated from aged garlic extract, exhibits broad-spectrum antioxidant activity by controlling reactive oxygen species generation and modulating the mitogen-activated protein kinase (MAPK). Reactive oxygen production was inhibited by NADPH oxidase 1, thus decreasing osteoclast fusion caused by the receptor activator of nuclear factor-kappa B ligand. Allicin, diallyl disulfide and diallyl trisulfide are the main antioxidant compounds that demonstrated an antioxidant effect at lower, physiologically relevant doses (Gruhlke et al., 2010). Abdel-Daim et al. (2018) reported that diallyl sulfide is a potent antioxidant with cytoprotective activites due to the inhibition of cytochrome P450-2E1, thereby reducing the formation of reactive oxygen and nitrogen species. Saponins extracted from garlic cloves were found to scavenge intracellular reactive oxygen species and protect mouse-derived C2C12 myblasts againts growth inhibition and H_2O_2-induced DNA damage (Shang et al., 2019). Meta-analysis done by Moosavian et al. (2020) shows that garlic supplementation improves antioxidant status by increasing total antioxidant capacity and decreasing malondialdehyde levels. Allicin, the major component of crushed garlic, can promote the phosphorylation of Sirtuin 1, upragulate the activity of $SIRT_1$, reduce the production of reactive oxygen species and decrease the expression of plasma activator inhibitor-1 in cells (Hu et al., 2016). Garlic sulfide can regulate mitochondrial respiration in cardiac miocytes and produce hydrogen sulfide via myocardial mitochondria, which can dilate contracted vascular smooth muscle (Li et al., 2016) and play the anti-myocardial ischemic effect. Garlic saponins exert antioxidant effects by inhibiting malondialdehyde content and restoring reduced superoxide dismutase activity (Miao et al., 2020). In a study on New Zealand rabbits receiving a high-cholestrol diet, treatment with allicin significantly reduced malondialdehyde, superoxide dismutase and glutathione levels (El-Sheakh et al., 2016). The antiatherogenic effect of allicin may be related to its ability to scavenge free radicals and restore antioxidant defense systems.

4.6 GARLIC, BRAIN-DERIVED NEUROTROPHIC FACTOR AND NEUROPROTECTION

Brain-derived neurotrophic factor (BDNF) is a small-sized neurotrophin, which plays a critical role in the growth and development of neurons (Mizui and Kojima, 2018). It has been shown to influence non-neuronal cells and their function in neuronal development. Liver, heart and lungs express BDNF in smaller amounts (Bathina and Das, 2015). Urbán and Guillemot (2014) reported that neurogensis usually occurs exclusively during embryonic development; newly produced hippocampal cells can be incorporated into neural networks during adolescence. Hippocampal neurogensis is important for the generation of new synapses and preservation of old ones. Various impulses, which directly affect netrotransmitters, neurotrophic and growth factor, as well as environmental impulses, can control neurogenesis in adults (Zhao et al., 2008; Liu and Nusslock, 2018). Feeding of certain doses of garlic to diabetic rats caused memory upgradation, an increase in hippocampul Ca^{2+} ATPase activity and glutamine synthetase activity (Semuyaba et al., 2017). A number of studies have been done on the protective performance of fresh garlic or aged garlic extract in neurodegenenerative conditions like Alzheimer's disease and cereberal ischemia (Chauhan, 2005; Qu Z et al., 2016; Song et al., 2020; Tedeschi et al., 2022). The neuroprotective effect of garlic might be due to the boosting of tissue immunity against oxidative stress, which decreases free redicals such as NO_2 and O_2, while also increasing mitochondrial function (Jeremic et al., 2019; He et al., 2021; Leurcharusmee et al., 2021).

In an experimental study done by Colin et al. (2011), rats were subjected to ischemia for 60 minutes followed by a 24-hour reperfusion phase. These rats were supplemented with aged garlic extract (1.2 ml/kg weight), which resulted reduced TNF and COX-2 protein levels and function. This indicates that aged garlic extract's neuroprotective qualities are associated with both its antioxidant characterstics and its ability to reduce TNF concentrations and COX-2 protein production and function. Aged garlic extract could reduce the damage cused by neuronal injury (Colin et al., 2011).

In another study done by Wang et al. (2020), diallyl disulfide (25 to 50 mg/kg) and diallyl trisulfide (20 to 40 mg/kg) were supplemented to chronic constriction injury rats for 14 days. H_2S, brain-derived neurotrophic factor and nuclear factor erythroid 2-related factor 2(Nrf2) amounts in the sciatic nerve and dorsal root ganglia were restored after administration of these organosulfur chemicals. These experiments clearly indicate that diallyl disulfide and diallyl trisulfide therapy caused a significant reduction in neuropathic pain in chronic constriction injury-treated rats. Another garlic component, allixin (1–100 mg/ml), added to the cell environment could significantly increase the viability of nerve cells obtained from different parts of the brain (Moriguchi et al., 1997).

Jung et al. (2016) found that oral administration of garlic oil (10 ml/kg) to mice once a day for three weeks caused an elevation in brain-derived neurotrophic factor concentration and a decrease in acetylcholinesterase (AChE) function in hippocampal homogenate. Garlic oil could lower the frontal cortex recycling ratio of neurotransmitters like serotonin and dopamine, raising their levels (Huang et al., 2019). Long-term administration of garlic oil can enhance hippocampal brain-derived neurotrophic factor via monoamine neurotransmitter regulation as well as their associated signaling cycle (Jung et al., 2016).

Ji et al. (2013) reported some contradictory findings on the effect of sulfur components of garlic. According to them, diallyl sulfide significantly downregulated the proliferation of nerve cells. Treatment with 10 mg/kg diallyl disulfide decreased hippocampal brain-derived neurotrophic factor concentrations, which ultimately reduced the neurotrophic factor signaling, phosphorylated CREB signaling and phosphorylated ERKs—all of which are linked to neural stem cell proliferation and differentiation in the hippocampus. They found that diallyl disulfide might have negative consequences on neural cell proliferation in the hippocampus (Ji et al., 2013).

Alzheimer's disease is the major cause of dementia in elderly human beings and leads to neural disorders (Kumar and Kumar, 2003). The acetylcholine (ACh) system is involved in the pathophysiology noted in Alzheimer's disease. Acetylcholinesterase (AChE) is the main enzyme that hydrolyzes ACh into acetate and choline (Singh and Singh, 1996b). Donepezil (an AChE inhibitor) has been effective in the management of Alzheimer's disease. Garlic oil suppresses AChE activity in cerebral cortex synaptosomes and exhibits antioxidant properties (Akinyemi et al., 2018). Thus, this anti-AChE activity, as well as their ability to scavenge diphenyl-1-picrylhydrazyl (DPPH) free radicals to assess the antioxidant activity and reduce Fe^{3+} to Fe^{2+}, could be suggested as possible mechanisms for the neuroprotective potential of garlic (Piscitelli et al., 2002).

Borek (2006) demonstrated that the neuroprotective effect of aged garlic extract helps protect the brain from neurodegenerative diseases by preventing brain injury following ischemia, saving neurons from apoptosis and inhibiting oxidative death caused by β-amyloid (Jackson et al., 2003). Aggregated β-amyloid accumulation in the brain is believed to be one of the pathological factors driving the onset of Alzheimer's disease. According to Haider et al. (2008), prolonged garlic intake is related to promoting memory function by affecting the levels of the neurotransmitter serotonin. Consumption of garlic extracts has been shown to improve memory by eliminating free redicals that cause oxidative damage and by inhibiting the AChE enzyme (Mukherjee and Banerjee, 2013). Kumar (2015) found that allicin inhibits AChE and butyrylcholinesterase (BuChE) enzymes, which increases ACh concentration in the brain. AChE inhibitors could be part of a combination therapy against Alzheimer's disease (Schmitt et al., 2004). Allicin is a small lipophilic molecule that can suppress BuChE and AChE concentrations in patient's brain (Inglis, 2002). Liu et al. (2015) found that allicin has a protective effect on ischemic neuronal damage controlled by apoptosis and oxidative stress pathways.

REFERENCES

Abdel-Daim MM, Shaheen HM, Abushouk AI, Toraih EA, Fawzy MS, Alansari WS, Aleya L, Bangau S (2018) Thymoquinone and dially sulfide protect against fipronil-induced oxidative injury in rats. *Environ Sci Pollut Res Int*. 25: 23909–23916.

Abdullah TH, Kandil O, Elakdi A, Carter J (1988) Garlic revisited: Therapeutic for the major diseases of our times. *J Natl Med Assoc*. 80: 439–445.

Adetumbi MA, Javor GT, Lau BHS (1986) *Allium sativum* (garlic) inhibits lipid synthesis by *Candida albicans*. Antimicrobial agents. *Chemother*. 30: 499–501.

Adetumbi MA, Lau BHS (1983) *Allium sativum* (garlic): A natural antibiotic. *Med Hypothesis.* 12: 227–237.

Afarid M, Sadeghi E, Johari M, Namvar E, Sanie-Jahromi F (2022) Evaluation of the effect of garlic tablet as a complementary treatment for patients with diabetic retinopathy. *J Diabetes Res.* 2022: 1–7. http://doi.org/10.1155/2022/6620661.

Afkhami-Ardekani M, Kamali-Ardekani A-R, Ahmad S (2006) Effect of garlic on serum lipids and blood glucose of type 2 diabetic patients. *Int J Diabetes Dev Countries.* 26(2): 86–88. https://doi.org/10.4103/0973-3930.28279.

Agarwal KC (1996) Therapeutic actions of garlic constituents. *Med Res Rev.* 16: 111–124.

Agarwal MK, Iqbal M, Athar M (2007) Garlic oil ameliorates ferric nitrilotriacetate (Fe-NTA)—induced damage and tumor promotion: Implications for cancer prevention. *Food Chem Toxicol.* 45: 1634–1640.

Ahsan M, Islam SN (1996) Garlic: A broad spectrum antibacterial agent effective against common pathogenic bacterial. *Fitoterapia.* 67: 374–376.

Ajamzibad H, Beydokhti TB, Mohtasham F, Nematollahi MR (2021) Effect of boiled garlic and shirazi lemon on blood lipids in hyperlipidemic patients: A quasi-experimental study. *J Herb Med.* 28: 100439 https://doi.org/10.1016/j.hermed.2021.100439.

Akinyemi AJ, Le Kan Faboya AP, AwOnegan AP, ISrael O, Anadozie S, Titus AO (2018) Antioxidant and anti-acetylcholinesterase activities of essential oils from garlic (*Allium sativum*) bulbs. *Int J Plant Res.* 31(2). https://doi.org/10.4172/2229-4473.1000397.

Alam K, Hoq O, Uddin S (2016) Medicinal plant *Allium sativum* a review. *J Med Plants Stud.* 4(6): 72–79.

Al-Brakati AY (2016) Effect of garlic against diabetic retinopathy in adult albino rats. *Res J Pharma Biol Chem Sci.* 7(5): 2748–2759.

Allison GL, Lowe GM, Rahman K (2012) Aged garlic extract inhibits platelet activation by increasing intra cellular CAMP and reducing the interation of GPIIb/IIIa receptor with fibrinogen. *Life Sci.* 91: 1275–1280.

Al-Waili Saloom KY, Akmal M, Ali-Waili H, Ali A, Al-Sahani K (2007) Effects of heating/storage and ultraviolet exposure on antimicrobial activity of garlic juice. *J Med Food.* 10: 208–212.

Amer M, Taha M, Tassan Z (1980) The effect of aqueous garlic extract on the growth of dermatophytes. *Int J Dermatol.* 19: 285–287.

Andrianova IV, Fomchenkov IV, Orekhov AN (2002) Hypotensive effect of long acting garlic tablets allicor (a double—blind placebo—controlled trial). *Ter Arkh.* 74: 76–88.

Ankri T, Rabinkov A, Wilchek M, Mirelman D (1997) Allicin from garlic strongly inhibits cysteine proteinases and cytoplasmic effects of *Entamoeba histolytica. Antimicrob Agents Chemother.* 41: 2286–2288.

Anonymous (1986) Zheziang Institute of Traditional Chinese Medicine, The effect of essential oil of garlic on hyperlipidemia and platelet aggregation. *J Tradit Chin Med.* 6: 117–120.

Apitz-Castro R, Cabrea S, Cruz MR, LEdezma E, Jain MK (1983) Effect of garlic extract and of three pure components isolated from it on human platelet aggregation, archidonate metabolism, release reactivity and platelet ultrastructure. *Thromb Res.* 32: 155–169.

Apitz-Castro R, Escalanate J, Vargas R, Jain MK (1986) Ajone, the antiplatelet principle of garlic, synergistically potentiates the antiaggregatory action of prostacyclin, for scolin, indomethiacin and dypiridamole on human platelets. *Thromb Res.* 42: 303–311.

Appleton JA, Tansey MR (1975) Inhibition of growth of zoopathogenic fungi by garlic extract. *Mycologia.* 67: 882–885.

Aqel MB, Gharaibah MN, Salhab AS (1991) Direct relaxant effects of garlic Juice on smooth and cardiac muscles. *J Ethanopharmacol.* 33: 13–19.

Aquilano K, Filomeni G, Baldelli S, Piccirillo S, De MArtino A, Rotilio G, Ciriolo MR (2007) Neuronal nitric oxide synthase protects neuroblastoma cells from oxidative stress mediated by garlic derivatives. *J Neurochem.* 101: 1327–1337.

Arditti FD, Rabinkov A, Miron T, Reisner Y, Berrebi A, Wilchek M, Mirelman D (2005) Apoptotic killing of B-chronic lymphocytic leukemia tumor cells by allicin generated *in Situ* using a rituximab-alliinase conjugate. *Mol Cancer Ther.* 4: 325–331.

Ariga T, Seki T (2006) Antithrombotic and anticancer effects of garlic derived sulfur compounds: A review. *Biofactors.* 26: 93–103.

Asdaq SM, Inamdar MN (2010) Potential of garlic and its active constituents, S-allyl cysteine, as antihypertensive and cardioprotectiev in presence of captoril. *Phytomedicine.* 17: 1016–1026.

Asdaq SM, Inamdar MN (2011a) The potential benefits of a garlic and hydrochlorothiazide combination as antihypertensive and cardioprotective in rats. *J Nat Med.* 65: 81–88.

Asdaq SMB, Inamdar MN (2011b) Pharmacodynamic and pharmacokinetic interactions of propanol with garlic (*Allium sativum*) in rats. *Evid Based Complement Alternat Med.* 2011: 824042. http://doi.org/10.1093/ecam/neq076.

Ashraf R, Aamir K, Shaikh AR, Ahmed T (2005) Effects of garlic on dyslipidemia in patients with type 2 diabetes mellitus. *J Ayub Med Coll Abbottabad.* 17(3): 60–64.

Atkin M, Laight D, Cummings MH (2016) The effects of garlic extract upon endothelial function, vascular inflammation, oxidative stress and insulin resistance in adults with type 2 diabetes at high cardiovascular risk. A pilot double blind randomized placebo controlled trial. *J Diabetes Complications.* 30(4): 723–727.

Augusti KT (1977) Hypocholesterolemic effects of garlic (*Allium sativum* Linn.). *Indian J Exp Biol.* 15: 489–490.

Augusti KT (1996) Therapeutic values of onion (*Allium cepa* L.) and garlic (*Allium satium* L.). *Indian J Exp Biol.* 34: 634–640.

Augusti KT, Mathew PT (1974) Lipid lowering effect of allicin (diallyl disulphide-oxide) on long term feeding to normal rats. *Experientia.* 30: 468–470.

Badol P, David-Dufilho M, Anger J, Whiteheart SW, Rendn F (2007) Thiosulfinates modulates platelet activation by reaction with surface free sulfhydryls and internal thiol-containing proteins. *Platelets.* 18: 481–490.

Bagul M, Kakumanu S, Wilson TA (2015) Crude garlic extract inhibits cell proliferation and induces cell cycle arrest and apoptosis of cancer cells *in Vitro. J Med Food.* 18: 731–737. https://doi.org/10.1089/jmf.214.0064.

Bakhshi M, Taher JB, Shabestari SB, Tanik A, Pahlevan R (2012) Comparison of therapeutic effects of aqueous extract of garlic and nystain mouthwash in denature stomatitis. *Gerodontology.* 29: 680–684.

Bakri IM, Douglas CW (2005) Inhibitory effect of garlic extract on oral bacteria. *Arch Oral Biol.* 50: 645–651.

Baliga M, Shivashankara A, Palatty P, Dsouza J, Arora R (2013) Chapter 34: Protective effect of garlic (*Allium sativum* L.) against atherosclerosis A2. In *Bioactive Food as Dietary Interventions for Cardiovascular Disease.* Preedy RR, Watson VR, Eds. Academic Press, San Diego, pp. 591–607.

Banerjee SK, Maulik SK (2002) Effect of garlic on cardiovascular disorders: A review. *Nutr J.* 1: 4.

Barankovic S, RaDenkovic M, Kitic D, Veljkovic S, Ivetic V, Pavlovi D (2011) Comparison of hypotensive and bradycardic activity of ginko, garlic and onion extracts. *Clin Exp Hypertens.* 33: 95–99.

Barone FE, Tansey MK (1977) Isolation, purification, identification, synthesis and kinetics of activity of the anti candidal component of *Allium sativum* and a hypothesis for its mode of action. *Mycologia.* 69: 793–825.

Bathina S, Das UN (2015) Brain derived neurotrophic factor and its clinical implications. *Arch Med Res.* 11(6): 1164–1178.

Batiha GE-S, Beshbishy AM, Wasef LG, Elewa YHA, AI-Sagen AA, El-Hack MEA, Taha AE, Abd-Elhakim YM, Devkota HP (2020) Chemical constituents and pharmacological activities of garlic (*Allium sativum* L.): A review. *Nutrients.* 12: 872. https://doi.org/10.3390/nu/2030872.

Bazaraliyeva A, Moldashov D, Turgumbayeva A, Kartbayeva E, Kalykova A, Sarsenova L, Issayeva R (2022) Chemical and biological properties of bioactive compounds from garlic (*Allium sativum*). *Pharmacia.* 69(4): 955–964.

Belman S (1985) Inhibition of soybean lipoxygenase by onion and garlic oil constituents. *Proc Am Assoc Cancer Res.* 26: 131.

Benavides GL, Squadrito GL, Mills RW, Patel HD, Isbell TS, Patel RP, Darley-Usmar VM, Doeller JE, Kraus DW (2007) Hydrogen sulfide mediates the vasoactivity of garlic. *Proc Natl Acad Sci USA.* 104: 17977–17982.

Bespalov VG, Nlu B, Ivanova OA, Krzhivitskii PI, Semiglazov VF, Aleksandrov VA, Sobenin NA, Orekhov AN (2004) Study of an antioxidant dietary supplement Karinat in patients with benign breast disease. *Vopr Onkol.* 50(4): 467–472.

Bhattacharya S, Maji U, Khan GA, Das R, Sinha AK, Ghosh C, Maiti S (2019) Antidiabetic role of a novel protein from garlic via NO in expression of Glut-4/insulin in liver of alloxan induced diabetic mice. Biomed Pharmacother. 111: 1302–1314.

Bhushan S, Sharma SP, Singh SP, Agarwal S, Indraycin A, Seth P (1979) Study of the hypocholestrolemic effect of onion (*Allium cepa*) on normal human beings. *Indian Med Gaz.* 16: 249–251.

Blahova J, Martiniakova M, Babikova M, Kovacova V, Mondockova V, Omelka R (2021) Pharmaceutical drugs and natural therapeutic products for the treatment of type 2 diabetes mellitus. *Pharmaceuticals.* 14(8): 806.

Blanch M, Carrom M, Ranilla M, Viso A, Vazquez-Anon M, Bach A (2016) Influence of a mixture of cinnamaldehyde and garlic oil on rumen fermentation, feeding behavior and performance of lactating dairy cows. *Anim Feed Sci Technol.* 219: 313–323.

Block E (1985). The chemistry of garlic and onions. *Sci Am.* 252: 114–119.

Block E (1998) Garlic as a functional food: A status report. In *Functional Foods for Disease Prevention.* Shibamoto T, Terao J, Qsawa T, Eds, Vol. 2. American Chemical Society, Washington, pp. 125–143.

Block E, Iyer R, Grisoni S, Saha C, Belman S, Lossing FP (1988) Lypooxygenase inhibitors from the essential oil of garlic. Markovinikov addition of the allyldithio radical to olefins. *J Am Chem Soc.* 110: 7813–7827.

Boivin D, Lany S, Lord-Dufours, Jackson J, Beauliew E, Cote M, Moghrabi A, Barrette S, Gingra SD, Belivean R (2009) Antiproliferative and antioxidant activities of common vegetables: A comparative study. *Food Chem.* 112: 374–380.

Bolton S, Null G, Troetel WM (1982) The medical uses of garlic-fact and fiction. *Am Pharmacol.* 22: 40–43.

Bordia A (1981) Effect of garlic and blood lipids in patients with coronary heart disease. *Am Clin Nutr.* 34: 2100–2103.

Bordia A, Bansal HC (1973) Essential oil of garlic in prevention of atherosclerosis. *Lancet.* 2: 1491–1492.

Bordia AK, Joshi HK, Sandhya YK, Bhu N (1977) Effect of essential oil of garlic on serum fibrinolytic activity in patients with coronary heart disease. *Atherosclerosis.* 28: 155–159.

Bordia AK, Sodhya SK, Rathore AS, Bhu N (1978) Essential oil of garlic on blood llipids and fibrinolytic activity in patients with coronary arthery disease. *J Assoc Physicians India.* 26: 327–331.

Borek C (2001) Antioxidant health effects of aged garlic extract. *J Nutr Mar.* 131(3s): 1010s–1055s. https://doi.org/10.1093/jn/131.3.1010s.

Borek C (2006) Garlic reduces dementia and heart-disease risk. *J Nutr.* 136: 8105–8125.

Bui TPT, Tran TAM, Naguyen TTH, Le TH, Tran TH, Huynh TPL, Nguyen TT, Tran TVA, Phan TR, Phann VT, Nguyen VH, Duong TR, Nguyen TT, Vo TT, Lam KH, Nguyen TAN (2020) Investigation in to SARS-COV2 resistance of compound in garlic essential oil. *ACS Omega.* 5(14): 8312–8320.

Burz C, Berindan-Neagoe I, Balacescu O, Irimie A (2009) Apoptosis in cancer: Key molecular signaling pathway and therapy-targets. *Acta Oncol.* 48: 811–821. https://doi.org/10.1080/02841860902974175.

Canizares P, Gracia I, Gomez Argila C, Bioxeda D, de Rafael LDE (2004b) Allyl thiosulfinates, the bacteriostatic compounds of garlic against *Helicobacter pylori. Biotechnol Prog.* 20: 397–401.

Canizares P, Gracia I, Gomez LA, Garcia A, Martin de, Argila C, Bioxeda D, de Rafael L (2004a) Thermal degradation of allicin in garlic extracts and its implication on the inhibition of the *in vitro* growth of *Helicobacter pylori. Biotechnol Prog.* 20: 32–37.

Canugovi C, Stevenson MD, Vendrow AE, Hayami T, Robidou XJ, Xiao H, Zhang Y-Y, Eitzman DT, Runge MS, Madamanchi NR (2019) Increased mitochondrial NADPH oxidase 4 (NOX4) expressions in aging is causative factor in aortic stiffening. *Redox Biol.* 26(2019): 101288. https://doi.org/10.1016/j.redox.2019.101288.

Caporaso N, Smith SM, Ing RHK (1983) Antifungal activity in human urine and serum after injection of garlic antimicrobial agents. *Chemother.* 23: 700–702.

Catanzaro E, Canistro D, Pellicioni V, Vivarelli F, Fimognari C (2022) Anticancer potential of allicin: A review. *Pharmacol Res.* 177: 106118. https://doi.org/10.1016/J.phrs.2022.106118.

Cavagnaro PF, Camargo A, Galmarini CR, Simon PW (2007) Effect of cooking on garlic antiplatelet activity and thiosulfinates content. *J Agric Food Chem.* 55: 1280–1288.

Cavallito CJ, Bailey JH (1944a) Allicin: The antimicrobial principle of *Allium sativum* I. Isolation, Physical properties and antibacterial action. *J Am Chem Soc.* 66: 1950–1951.

Cavallito CJ, Bailey JH (1944b) Allicin: The antimicrobial principle of *Allium sativum* I. Isolation physical properties and antibacterial action. *J Am Chem Soc.* 66: 1952–1954.

Cavallito CJ, Buck JS, Suter CM (1944) Allicin: The antibacterial principle of *Allium sativum* II. Determination of chemical structure. *J Am Chem Soc.* 66: 1952–1954.

CFSAN (2007) *Center for Food Safety and Applied Nutrition of the U.S. Food and Drug Administration*, July. Available from URL: http://www.cfsan.fda.gov/~dms/hclmgui5.html

Chakraborty D, Majumder A (2020) Garlic (Lahsun)—an immunity booster against SARS—COV—2. *Biot Res Today.* 2(8): 755–757.

Chang HP, Chen YH (2005) Differential effects of organosulfur compounds from garlic oil on nitric oxide and prostaglandin E2 in stimulated macrophages. *Nutrition.* 21: 530–536.

Chang HS, Yamato O, Sakai Y, Yamasaki M, Maede Y (2004) Acceleration of superoxide generation in polymorphonuclear leukocytes and inhibition of platelet aggregation by alk(en)yl thiosulphates derived from onion and garlic in dogs and humans. *Prostaglandins Leukot Essent Fatty Acids.* 70: 77–83.

Chang MLW, Johnson MA (1980) Effect of garlic on carbohydrate metabolism and lipid synthesis in rats. *J Nutr.* 112: 931–938.

Charron CS, Dawson HD, Albaugh GP, Solverson PM, Vinyard BT, Solano-Aguilar GI, Molokin A, Novotny JA (2015) A single meal containing raw crushed garlic influence expression of immunity and cancer related genes in whole blood of humans. *J Nutr.* 145(11): 2448–2455.

Chauhan N (2005) Multiplicity of garlic health effects and Alzheimer's disease. *J Nutr Health Aging*. 9(6): 421–432.

Chavan RD, Shinde P, Girkor K, Madage R, Chowdhary A, (2016) Assessment of anti-influenza activity and hemagglutination inhibition of *Plumbago indica* and *Allium sativum* extracts. *Pharmacognosy Res*. 8: 105. https://doi.org/10.4103/0974-8490.172562.

Chen F, Li H, Wang Y, Gao M, Cheng Y, Liu D, Jia M, Zhang J (2016a) Inhibition of allicin in Eca/09 and EC9706 cells via G2/M phase arrest and mitochondrial apoptosis pathway. *J Funct Foods*. 25: 523–536. https://doi.org/10.1016/jff.2016.06.027.

Chen I, Li J, Luo H, Xn W, Chen G (2006) Binding interaction of quercetin-3–13 galactoside and its synthetic derivatives with SARS-COV 3CI. pro: Structure activity relationship studies reveal salient pharmacophore features. *Bioorg Med Chem*. 14: 8295–8306. https://doi.org/10.1016/j.banc.2006.09.014.

Chen Y, Sun J, Dou C, Li N, Kang F, Wang Y, Cao Z, Yang X, Dong S (2016b) Alliin attenuated RAKL-induced osteocalstogenesis by scavenging reactive oxygen species through inhibiting NOX1. *Int J Mol Sci*. 17(9): 1516. https://doi.org/10.3390/ijms17091516.

Chhabria SV, Akbarsha MA, Li AP, Kharkar PS, Desai KB (2015) *In Situ* allicin generation using targeted alllnase delivery for inhibition of MIA Pa Ca-2 cells via epigenetic changes, oxidative stress and cyclin-dependant kinase inhibitor (CDKI) expression. *Apoptosis*. 20: 1388–1409.

Cho SJ, Rhee DK, Pyo S (2006) Allicin a major component of garlic, inhibits apoptosis of macrophages in a depleted nutritional state. *Nutrition*. 22: 1177–1184.

Choudhary PR, Jani RD, Sharma MS (2018) Effect of raw crushed garlic (*Allium sativum* L.) on components of metabolic syndrome. *J Diet Suppl*. 15(4): 499–506.

Choudhury A, Siraj A, Biprojit P (2024) *Allium sativum*: Rediscovering its culinary and medicinal heritage. *J Appl Pharmacogn Phytochem*. 4(1): 20–26.

Chu Q, Ling MT, Feng H, Cheung HW, Tsao, SW, Wang X, Wong YC (2006) A novel anticancer effect of garlic derivatives: Inhibition of cancer cell invasion through restoration of E-cadherin expression. *Carcinogenesis*. 27: 2180–2189.

Chuah SC, Moore PK, Zhu YZ (2007) S-allycysteine mediates cardioprotection in an acute myocardial infarction rat model via a hydrogen sulfide-mediated pathway. *Am J Physiol Heart Circ Physiol*. 293(5): 2693–26701. http://doi.org/10.1152/ajpheart.00853.2007.

Chung MJ, Lee SH, Sung NJ (2002) Inhibitor effect of whole strawberries, garlic juice or kale juice on endogenous formation of N-nitrosodium methylamine in humans. *Cancer Lett*. 182: 1–10.

Colin-Gonzalez AL, Ortiz-Plata A, Villeda-Hernandez J, Barrera D, Molina-Jijon E, Pedraza-Chaverri J, Maldonado PD (2011) Aged garlic extract attenuates cerebral damage and cylooxygenase-2 induction after ischemia and reperfusion in rats. *Plant Food Hum Nutr*. 66(4): 348–354.

Coppi A, Cabinian M, Micelman D, Sinnis P (2006) Antimalarial activity of allicin, a biologically active compound from garlic cloves. *Antimicrob Agents Chemother*. 50: 1731–1737.

Curtiss LK, Black AS Takagi Y, Plone EF (1987) New mechanisms for foam cell generation in atherosclerotic lesions. *J Clin Invest*. 80: 367–373.

Dafriani P, Marlinda R, Arman E, Idaman M (2020) Garlic: An alternative in reducing blood glucose on diabetic patients. *Int J Community Med Public Health*. 7: 2078–2081.

Dahlback B (2000) Blood coagulation. *Lancet*. 355: 1627–1632. https://doi.org/10.1016/50140-6736(00)02225-X.

Damrau F (1941) The use of garlic concentrate in vascular hypertension. *Med Res*. 153: 249–251.

Danaei G, Finucane MM, Lu Y, Singh GM, Cowan MJ, Paciorek J, Lin JK, Farzadfar F, Khang YH, Stevens GA, Rao M, Ali MK, Rilly LM, Robinson CA, Ezzati M (2011)

National, regional and global trends in fasting plasma glucose and diabetes prevalence since 1980: Systematic analysis of health examination surveys and epidemiological studies with 370 country-years and 2–7 million participants. *Lancet*. 378(9785): 31–40.

Day HJ, Holmsen H, Zucker MB (1976) *Platelet Function Testing*. National Institute of Health Service, Bethesd (DHEW-NIH 78-1087).

De Greef D, Barton EM, Sandberg EN, Croley CR, Pumarol J, Wong, TL, Das N, Bishayee A (2021) Anticancer potential of garlic and its bioactive constituents: A systematic and comprehensive review. *Seminars Cancer Biol*. 73: 219–264. http://doi.org/10.1016/j.semcancer.2020.11.020.

Dini C, Fabbri A, Geraci A (2011) The potential role of garlic (*Allium sativum*) against the multidrug resistant tuberculosis pandemic: A review. *Ann Ist Super Sanita*. 47: 465–473.

Dirsch VM, Antlsperger DS, Hentze H, Vollmar AM (2002) Ajoene an experimental anti-leukemic drug: Mechanism of cell death. *Leukemia*. 16: 74–83.

Doll R (1977) Strategy for detection of cancer hazards to man. *Nature*. 265: 598–596.

Doll R, Peto R (1987) The cause of cancer: Quantitave extimation of avaoidable risk of cancer in the United States. *J Natl Cancer Inst*. 66: 1193–1198.

Dragos D, Manea MM, Timofte D, Ionescu D (2020) Mechanisms of herbal nephroprotection in diabetes mellitus. *J Diabetes Res*. 2020: 1–31.

Drobiova H, Thomson M, Al-Qattank, Peltonen-Shalaby R, Al-Amin Z, Ali-Amin Z, Ali-M (2011) Garlic increases antioxidant levels in diabetic and hypertensive rats determined by a modified peroxidase method. *Evid Based Complement Altern Med*. 703049. https://doi.org/10.1093/ecam/nep011.

Druesne PN, Pagniez A, Thomas M, Cherbuy C, Duee PH, Martel P, Chaumontet C (2006) Diallyl disulfide increases CDKN1A promoter-associated histone acetylation in human. *J Agric Food Chem*. 54: 7503–7507.

Dubey H, Singh A, Patole AM, Tenpe CR (2017) Antihypertensive effect of allicin in dexamethasone-induced hypertensive rats. *Integr Med Res*. 6: 60–65.

Ebid AIM, Hassan AG, Mohammed SA (2006) Aspirin, garlic and morbidity in patients with cardiovascular disorders: A prospective study. *Egyptian J Biomed Sci*. 22: 219–236.

Efendy JL, Simmons DL, Cambell GR, Campbell JH (1997) The effect of the aged garlic extracts "Kyolic" on the development of experimental atherosclerosis. *Atherosclerosis*. 132: 37–42.

Eidi A, Eidi M, Esmaeili E (2006) Antidiabitic effects of garlic (*Allium sativum* L.) in normal and streptozotocin induced diabetic rats. *Phytomedicine*. 13: 624–629.

Eilat S, Avramovitech D, Rabinkov A, Mirelman D, Battler A, Elder M, Vered Z (1996) Allicin: Hypolipidemic drugs. *Future*. 21: 1101–1104.

El-Bayoumy K, Sinha R, Pinto JT, Rivlin RS (2006) Cancer chemoprevention by garlic—containing sulfur and selenium compounds. *J Nutr*. 136 (35 Spl): 8645–8695.

Elkayam A, Mirelman D, Peleg E, Wilchek M, Miron T, Rabinkov A, Oron HM, Rosenthal T (2003) The effect of allicin on weight in fructose-induced hyperinsulinemic, hyperlipidemic, hypertensive rats. *Am J Hypertens*. 16: 1053–1056.

Elnima EI, Ahmed SA, Mekkawi AG, Mossa JS (1983) The antimicrobial activity of garlic and onion extracts. *Pharmazil*. 38: 747–748.

Elrod JW, Calvert JW, Morrison J, Doeller JE, Krans DW, Tao L, Jiao X, Scalia R, Kiss L, Szabo C, Kimura H, Chow, CW, Lefer DJ (2007) Hydrogen sulfide attenuates myocardial ischemia-reperfusion injury by preservation of mitochondrial function. *Natl Acad Sci USA*. 4: 15560–15565.

El-Saadony M, Ahmad MS, Sameh AK, Heba MS, Taia A Abd El-Mageed, Samar SA, Mohamed IE, Sara SE, Walid FAM, Ahmed EA, Betty TM, Noor A Albastaki, Aysha AA, Marawan K El-Tarabily, Synan FA, El-Tarabily KA, Salam AI (2024)

Garlic bioactive substances and their therapeutic applications for improving human health: A comprehensive review. *Front Immunol.* http://doi.org/10.3389/fimmu.2024.1277074.

El-Sabban F, Abouazra H (2008) Effect of garlic on altherosclerosis and its factor. *East Mediterr Health J.* 14: 195–205.

El-Sheakh AR, Ghoneim HA, Suddek GM, Anmar E (2016) Atlenuation of oxidative stress, inflammation and endothelial dysfunction in hypercholestrolemic rabbits by allicin. *Can J Physiol Pharmacol.* 94: 216–229. https://doi.org/10.1139/cjpp-2015-0267.

Ernst E, Weihmayr TH, Matrai A (1985) Garlic and blood lipids. *Br Med J (Cin Res Ed).* 291: 139.

Eser N, Yoldas A, Turk A, Kalayci YA, Yalcin A, Cicek M (2021) Ameliorative effects of garlic oil on FNDC5 and irisin sensitivity in liver of streptozotocin-induced diabetic rats. *J Pharm Pharmacol.* 73(6): 824–834.

Espirito S, Keiss HP, Meyer K, Buytenhek Roos Th, Dirsch V, Buniatian G, Ende C, Gunther J, Heise K, Kellert D, Lerche K, Pavlica S, Struck F, Usbeck E, Voigt S, Zellmer S, Princen JMG, Vollmar AM, Gebhardt R (2007) Garlic and cardiovascular diseases. *Med Aromat Plant Sci Biotechnol.* 1: 31–36.

Faradilla A, Rahman M (2020) The effect of single type black garlic (*Allium sativum* L.) extract on the cell of Langerhans islets and the kidney tubular microscopy in male wistar rats (*Rattus norvegicus*) models of diabetes mellitus. *Syst Rev Pharm.* 11(6): 290–296.

Fatima N, Baqri SSR, Alsulimani A, Fagoonee S, Slama P, Kesari KK, Roychoudhury S, Haque S (2021) Phytochemicals from Indian ethnomedicines: Promising prospects for the management of oxidants. 10(10): 1606. https://doi.org/10.3390/antiox10101606.

Ferri N, Yokoyama K, Sadilek M, Paoletti R, Apitz-Castro R, Gelb MH, Corsini A (2003) Ajoene, a garlic compound, inhibits protein prenylation and arterial smooth muscle cell proliferation. *Brit J Pharmacol.* 138: 811–818.

Fromtling RA, Bulmer GS (1978) *In vitro* effect of aqueous extract of garlic (*Allium sativum*) on the growth and viability of *Cryptococcus neoformans. Mycologia.* 70: 397–404.

Fufa B (2019) Antibacterial and antifungal properties of garlic extract (*Allium sativum*): A review. *Microbiol Res J Int.* 28: 1–5.

Fukao H, Hideki Y, Yoh-Ichi T (2007) Antithrombotic effects of odorless garlic powder both *in Vitro* and *in Vivo. Biosci Biotechnol Biochem.* 71(1): 84–90.

Fukao T, Hosono T, Satomi M, Seki T, Ariga T (2008) Chemoprotective effect of diallyl trisulfide from garlic against carbon tetra chloride induced acute liver injury of rats. *Bio Factors.* 21: 171–174.

Galeone C, Pelucchi C, Levi F, Negri E, Franceschi S, Talamini R, Giacosa A, La Vecchia C (2006) Onion and garlic use and human cancer. *Am J Clin Nutr.* 84: 1027–1032.

Gallwitz H, Bonse S, Martinez-Cruz A, Schlichting I, Schumacher K, Krauth-Siegal RL (1999) Ajoene is an inhibitor and subversive substrate of human glutathione reductose and *Trypanosoma cruzi* trypanothione reductase: Crystallographic, Kinetic and spectroscopic studies. *J Med Chem.* 42: 364–372.

Gamboa MR, Aranda-Gonzalz I, Mut-Martin M, Garcia-Miss MR, Dumonteil E (2007) *In Vivo* and *in Vitro* control of *Leislamania mexicana* due to garlic-induced NO production. *Scand J Immunol.* 66: 508–514.

Gamboa-León R, Paraguai-de-Souza E, Borja-Cabrera GP, Santos FN, Myashiro LM, Pinheiro Sousa CB (2006) Immunotherapy against visceral leishmaniasis with the nucleoside hydrolase-DNA vaccine of Leshmania donovani. *Vaccine.* 24: 4863–4873.

Gardner CD, Messina M, Lawson LD, Farquhar JW (2003) Say, garlic, ginkgo bilboba: Their potential role in cardiovascular diseases prevention and treatment. *Curr Atheroscler Rep.* 5: 468–475.

Gebhardt R (1993) Multiple inhibitory effects of garlic extracts on cholesterol biosynthesis in hepatocytes. *Lipids*. 28: 613–619.

Ghorbani A, Zarvandi M, Rakhshandeh HA (2019) A randomized controlled trial of a herbal compound for improving metabolic parameters in diabetic patients with incontrolled dyslipidemia. *Endocr Metab Immune Disord Drug*. 19(7): 1075–1082.

Goleone C, Pelucchi C, Dal Maso L, Negri E, Montella M, Zucchetto A, Talamini R, La Veecchia C (2008) Allium vegetables intake and endometrial cancer risk. *Public Health Nutr*. 12: 1576–1579.

Gonen A, Harats D, Rabinkov T, Miron D, Mirelman M, Wilchek L, Weiner E, Ulman H, Levkovitz H, Ben-Shushan D, Shaish A (2005) The antiatherogenic effect of allicin: Possible mode of action. *Pathobiology*. 72: 325–334.

Gonzalez CA, Riboli E (2006) Diet and cancer: Where we are, where we are going. *Nutr Cancer*. 56: 225–231.

Gorinstein S, Jastrzebski Z, Namiesnik J, Leontowicz H, Leontowicz M, Trask S (2007) The atherosclerotic heart disease and protecting properties of garlic: Contemporary data. *Mol Nutr Food Res*. 5: 1365–1381.

Goudappeala P, Gowda CY, Kashinath R (2021) Diallyl disulfide regulates purine metabolism and their metabolites in diabetes mellitus. *Int J Res Pharm Sci*. 65(1): 28–34.

Gouddappala P, Sukumar E, Kashinath R (2019) Influence of diallyl disulphide on hepatic gluconeogensis protein phosphorylation. *Int J Res Pharm Sci*. 10: 1327.

Greenbaum CJ (2002) Insulin resistance in type 1 diabeties. *Diabetes Metab Res Rev*. 18(3): 192–200. https://doi.org/10.1002/dmrr.291.

Groot PHE, Scheck LM (1984) Effects of fat ingestion on high density lipoprotein profiles in human sera. *J Lipid Res*. 4: 692.

Grover SP, Mackman N (2020) Tissue factor in atherosclerosis and atherothrombosis. *Atherosclerosis*. 307: 80–86 https://doi.org/10.1016/j.atherosclerosis.2020.06.003.

Gruhlke MC, Niccoc C, Batteux F, Slusareno AJ (2016) The effect of allicins, a reactive sulfur spices from garlic, on a selection of mammalian cell lines. *Antioxidants*. 6(1): 1. https://doi.org/10.3390/antiox-6010001.

Gruhlke MC, Nwachwukwu I, Arbach M, Anwar A, Noll U, Slusarenko AJ (2010) *Allicin from Garlic Effective in Controlling Several Plant Diseases, Is a Reactive Sulfur Species (RSS) That Pushes Cells into Apoptosis*. In Proceedings of the Modern Fungicides and Antifungal Compounds VI. 16th International Reinhardsbrunn Symposium, Friedrichrods, Germany, 25–29 April 1990.

Guo Y, Li ZX, Zhang JY, Ma JL, Zhang L, Zhang Y, Zhou T, Liu WD, Han ZX, Li WQ, Pan KF, You WC (2020) Association between lifestyle factors, vitamin and garlic supplementation and gastric cancer outcomes: A secondary analysis of a randomized clinical trial. *JAMA Netw Open*. 3(6): e206628. https://doi.org/10.1001/jamanentworkopen.2020.6628.

Hahm ER, Kim SH, Mathan SV, Singh RP, Singh SV (2021) Mechanistic targets of diallyl trisulfide in human breast cancer cells identified by RNA-Seq analysis. *J Cancer Prev*. 26(2): 128–136. https://doi.org/10.15430/JCP.2021.26.2.128.

Haider S, Naz N, Khaliq S, Perveen T, Haleem DJ (2008) Repeated administration of fresh garlic increases memory retention in rats. *J Med Food*. 11: 675–679.

Hall HO (1913) The onion (*Allium cepa*) and garlic (*Allium sativa*) as a remedy for pneumonia and pulmonary tuberculosis. *Am Med*. 19: 26–34.

Harenberg J, Giese C, Zimmermann R (1988) Effect of dried garlic on blood coagulation, fibrinolysis, platelet aggregation and serum cholesterol levels in patient with hyperlipoprotenemia. *Atherosclerosis*. 74: 247–249.

Harris JC, Cottrell SL, Plummer S, Lloyed D (2001) Antimicrobial properties of *Allium sativum* (garlic). *Appl Microbial Biotechnol*. 57: 282–286.

Harris JC, Plummer S, Turner MP, Lloyel D (2000) The microaerophilic flagellati *Giardia intestinalis*: *Allium sativum* (garlic) is an effective antigiardial. *Microbiology*. 146: 3119–3127.

Hartwell JL (1967) Plants used against cancer: A survey. *Lloydia*. 30: 379–436.

Hartwell JL (1968) Plants used against cancer: A survey. *Lloydia*. 31: 71–170.

Hassan HT (2004) Ajoene (natural garlic compound): A new anti-leukaemia agent for AML therapy. *Leukemia Res*. 28: 667–671.

Hazaa IKK, Al-Taai NA, Khalil NK, Zakri AMM (2016) Efficacy of garlic and onion oils on murin experimental *Cryptosporidium parvum* infection. *Al-Anbar J Vet Sci*. 9: 69–74.

He H, Ma Y, Huang H, Huang C, Chen Z, Chen D, Guy, Wang X, Chen J (2021) A comprehensive understanding about the pharmacological effect of diallyl disulfide other than its anti-carcinogenic activities. *Eur J Pharmacol*. 893: 173803. http://doi.org/10.1016/j.ejphar.2020.173803.

Herman-Antosiewicz A, Powolny A, Singh SV (2007) Molecular targets of cancer chemoprevention by garlic-derived organosulfides. *Acto Pharmacol Sinica*. 28: 1355–1364.

Herman-Antosiewicz A, Singh SV (2004) Signal transduction pathways leading to cell cycle arrest and apoptosis induction in cancer cells by *Allium* vegetable-derived organosulfur compounds: A review. *Mutation Res*. 555: 121–131.

Hirsch K, Danilenko M, Giat J, Miran T, RAbinkov A, Wilchek M, Mirelman D, Levy J, Sharoni Y (2000) Effect of purified allicin, the major ingredient of freshly crushed garlic on cancer cell proliferation. *Nutr Cancer*. 38(2): 245–254. http://doi.org/10.1207/s15327914NC382_14.

Hiyasat B, Sabha D, Grotzinger K, Kempfert J, Rauwald JW, Mohr FW, Dhein S (2009) Antiplatelet activity of *Allium ursinum* and *Allium sativum*. *Pharmacology*. 83: 197–204.

Hodge G, Davis S, Rice M, Tapp H, Saxon B, Revesz T (2008) Garlic compounds selectively kill childhood pre-β acute lymphoblastic leukemia cells *in Vitro* without reducing T-cell function: Potential therapeutic use in the treatment of ALL. *Biologics*. 1: 143–149.

Houshmand B, Mahjour F, Dianat O (2013) Antibacterial effect of different concentrations of garlic (*Allium sativum*) extract on dental plaque bacteria. *Indian J Den Res*. 24–71.

Hu H, Pan Y, Fan X, Hu X, Zou W, Lin X (2016) Allicin inhibits H2O2-induced senescence in human umbilical vein endothelial cells through activation of Sirt1. *Chin J Biochem Mol Biol*. 32: 536–543. https://doi.org/10.13865/j-cnki.cjbmb.2016.05.09.

Huang J, Yang B, Xiang T, Peng W, Qin Z, Wan J, Zhang Li, Li H, Ren G (2015) Diallyl disulfide inhibits growth and metastatic potential of human triple-negative breast cancer cells through inactivation of the β-catenin signaling pathway. *Mol Nutr Food Res*. 59(6): 1063–1075. https://doi.org/10.1002/mnfr.201400668.

Huang Y-J, Lu K-H, Lin Y-E, Panyod S, Wu H-Y, Chang W-T, Sheen L-Y (2019) Garlic essential oil mediates acute and chronic mild stress-induced depression in rats via modulation of monoaminergic neurotransmission and brain-derived neurotrophic factor levels. *Food Funct*. 10(1): 8994–8105.

Hughes BG, Lawson LD (1991) Antimicrobial effects of *Allium sativum* L. (garlic), *Allium ampiloprsum* L., elephant garlic and *Allium cepa* L. (onion) compounds and commercial garlic supplement products. *Phytother Res*. 5: 154–158.

Ichikawa M, Ryu K, Yoshida J, Ide N, Kodera Y, Sasaoka T, Rosen RT (2003) Identification of six phenylpropanoids from garlic skin as major antioxidants. *J Agric Food Chem*. 51: 7313–7317.

Iciek M, Kwiecien I, Chwatko G, Sokolowska-Jezewicz M, Kowalczyk-Pachel D, Rokita H (2012) The effects of garlic–derived sulfur compounds on cell proliferation caspase 3 activity, thiol levels and anaerobic sulfur metabolism in human hepatoblastoma Hep G2 cells. *Cell Biochem Funct*. 30: 198–204.

Ide N, Nelson AB, Lau BHH (1997) Aged garlic extract and its constituents inhibit CU-2+ induced oxidative modification of low density lipoprotein. *Plasma Med.* 63: 263–264.

Inglis F (2002) The tolerability and safety of cholinesterase inhibitors in the treatment of dementia. *Int J Clin Pract Suppl.* 127: 45–63.

Irfan M, Kim M, Kwon H-W, Ree MH, Kim H-K (2018) Antiplatelet effects of garlic and chitosan: A comparative study between fermented and non-fermented preperations. *Biomed Sci Lett.* 24: 280–284. http://doi.org/10.15616/BSL.2018.24.3.280.

Isaacsolin JL, Moser M, Stein EA, Dudley K, Davey JA. Liskov E (1998) Garlic powder and plasma lipids and lipoproteins: A multi center, randomized, placebo controlled trial. *Arch Inter Med.* 158: 1189–1194.

Isensee H, Retz B, Jacob R (1993) Cardioprotective actions of garlic (*Allium sativum*). *Arztl Forsch.* 43: 94–58.

Islam MS, Choi H (2008) Comparative effects of dietary ginger (*Zingiber officinale*) and garlic (*Allium sativum*) investigated in a type 2 diabetes model of rats. *J Med Food.* 11(1): 152–159.

Iweala EE, Akubugwo EI, Okeke CV (2005) Effect of ethanolic extracts of *Allium sativum* Linn. Liliaceae on serum cholesterol and blood sugar levels of alibino rabbits. *Plant Prod Res J.* 9: 14–18.

Jackson R, McNeil B, Taylor C, Holl G, Ruff D, Gwebu E (2003) Effect of aged garlic extract on human recombinant caspace-3activity. *J Ala Acad Sci.* 74: 121–122.

Jacob C, Anwar A, Burkholz T (2008) Perspective on recent developments on sulfur-containing agents and hydrogen sulfide signaling. *Planta Med.* 74: 1580–1592.

Jain MK, Apitz-castro R (1987) Garlic: Molecular basis of the putative "Vampire repellent" action and other matters related to the heart and blood. *Trends Biochem Sci.* 12: 252–254.

Jain RC (1977) Effect of garlic on serum lipids, coagulability and fibrinolytic activity. *Am J Chin Nutr.* 30: 1380–1381.

Jain RC, Andleigh HS (1969) Onion and blood firinolytic activity. *Br Med J.* 1(5642): 514.

Jang HJ, Lee HJ, Yoon DK, Ji DS, Kim J, Lee CH (2018) Antioxidant and antimicrobial activities of fresh garlic and aged garlic by-products extracted with different solvents. *Food Sci Biotechnol.* 27(1): 219–225.

Jang HJ, Lee HJ, Yoon DK, Ji DS, Kim JH, Lee CH (2017) Antioxidant and antimicrobial activities of fresh garlic and aged garlic by-products extracted with different solvents. *Food Sci Biotechnol.* 27: 219–225.

Jeremic JN, Jakovljevic VL, Zivkovic VI, Srejovic IM, Bradic JV, Bolevich S, Nikalic Turnic TR, Mitrovic SL, Jovicic NU, Tyagi SC, Jeremic NS (2019) The cardioprotective effects of dially trisulfide on diabetic rats with ex vivo induced ischemia/reperfusion injury. *Mol Cell Biochem.* 460(1): 151–164.

Jezpwa J, Rafinski T, Wrocinski T (1966) Badania and dzialaniem antybiotycznym Czosnkv pospolitego (*Allium sativum* L.). *Herba Pol.* 12: 66.

Ji ST, Kim MS, Park HR, LEe E, Lee Y, Jang YJ, Kim HS, Lee J (2013) Diallyl disulfide impairs hippocampal neurogenesis in the young adult brain. *Toxicol Lett.* 221(1): 31–38.

Jiang X, Zhu X, Huang W, Xu H, Zhao Z, Li S, Lishanzhong, Cai J, Cao J (2017) Garlic-deerived organosulfur compound exerts antitumor efficacy via activation MAPK pathway and modulation of cytokines in SGC-7901 tumor-bearing mice. *Int Immunopharmacol.* 48: 135–145. https://doi.org/10.1016/j.indimp.2017.05.004.

JiKihara H, Qi G, Nozoe K, Hirokawa M, Sato H, Sugihara Y, Shimamoto F (2015) Aged garlic extract inhibits 1–2-dimethyl hydrazine-induced colon tumor development by suppressing cell proliferation. *Oncal Rep.* 33(3): 1131–1140.

Johnson MG, Vaughn RH (1969) Death of *Salmonella typhimurium* and Escheichia colic in the presence of freshly reconstituted dehydrated garlic and onion. *Appl Microbiol.* 17: 903–906.

Jonkers D, Broek EVD, Dooren IV, Thijs C, Dorant E, Hageman G, Stobberingh E (1999) Antibacterial effect of garlic and omeprazole on *Helicobacter pylori*. *J Antimicrobial Chemother*. 43(6): 837–839.

Josling P (2001) Preventing the common cold with a garlic supplement: A double-blind, placebo-controlled survey. *Adv Ther*. 18: 189–193.

Jung EM, Jung F, Mrowietz C, Kiesewetter H, Pundir G, Wenzel E (1991) Influence of garlic powder on cutaneous microcirculation. *Art Forschung*. 41: 626–630.

Jung F, Wolf S, Kisewetter H, Mrowietz C, Pundir G, Heiden M, Miayashita C, Wenzel E, Reim N (1989) Wirking von knoblaush anf Ergebnisse placebo on trollierter Pilotstudien an gesunden Probanden. *Nat Ganzheismed*. 7: 216–221.

Jung HY, Lee KY, Yoo DY, Kim JW, Yo M, Lee S, Yoo KY, Yoon YS, Choi JH, Hwang IK (2016) Essential oils from two *Allium* species exerts effects on cell proliferation and neuroblast differentiation in the mouse dentate gyrus by modulating brain-derived neurotrophic factor and acetylcholinesterase. *BMC Complement Altern Med*. 16(1): 431. http://doi.org/10.1186/s12906-016-138-4-6.

Karuppiah P, Rajaram S (2012) Antibacterial effect of *Allium sativum* cloves and *Zingiber officinale* rhizomes against multiple-drug resistant clinical pathogens. *Asian Pac J Trop Biomed*. 2(8): 597–607.

Kasuga S, Ushijima M, Morihara N, Itakura Y, Nakata Y (1999) Effect of aged garlic extract (AGE) on hyperglycemia induced by immobilization stress in mice. *Nihon Yakurigaku Zasshi*. 114(3): 191–197.

Kendler BS (1987) Garlic (*Allium sativum*) and onion (*Allium cepa*): A review of their relationship to cardiovascular diseases. *Pre Med*. 16: 670–685.

Khalid FA, Abdalla NM, Mohomed HEO, Toum AM, Magzoub MMA, Ali MS (2005) *In Vitro* assessment of anti-cutaneous leishmaniasis activity of some sulanese plants. *Acto Parasitol Turcica*. 29: 3–6.

Khan KI, Khan FZ, Nazar S (1985) The antimicrobial activity at *Allium sativum* (garlic), *Allium cepa* (onion) and *Raphanus sativus* (radish). *J Pharm Pb Univ Lhr*. 6: 59–71.

Khanum F, Anilakumar KR, Viswanathan KR (2004) Anticarcinogenic properties of garlic: A review. *Crit Rev Food Sci Nutr*. 44: 479–488.

Khedekar SS, Tushar GK, Saad MJK, Aniket GK, Kushwaha A, Sonali V (2024) Garlic (*Allium sativum*) its phytochemistry, pharmacognosy & medicinal uses. *Int J Res Publ Rev*. 5(1): 2624–2632.

Khubber S, Hashemifesharaki R, Mohammadi M, Gharibzahedi SMT (2020) Garlic (*Allium sativum* L.), a potential unique therapeutic food rich in organosulfur and flavonoid compounds to fight with COVID-19. *Nutr J*. 19: 124. https://doi.org/10.1186/s12937-020-00643-8.

Kiesel VA, Stan SD (2017) Diallyl trisulfide, a chemopreventive agent from *Allium* vegetables inhibits alpha-secretases in breast cancer cells. *Biochem Biophysical Res Commun*. 484(4): 833–838. https://doi.org/10.1002/mc.22364.

Kiesewetter H, Jung F, Pundir G, Jung EM, Mrowietz C, Wenzel E (1991) Effect of garlic on thromboiyte aggregation, microcirculation and risk factors. *Int J Clin Pharmacol Ther Toxicol*. 29: 151–155.

Kim JW, Kim YS, Kyung KH (2004) Inhibitory activity of essential oils of garlic and onion against bacteria and yeasts. *J Food Prot*. 67: 499–504.

Kirkman MS, McCarren M, Shah J, Duckworth W, Abraira C, Group VS (2006) The association between metabolic control and prevalent macrovascular disease in type 2 diabetes: The VA cooperative study in diabetes. *J Diabetes Complications*. 20(2): 75–80.

Kobiyama K, Ley K (2018). Atherosclerosis: A chronic inflammatory disease with an atommune. *Circ Res*. 123: 1188–1120. https://doi.org/10.1161/CIRCRESAHA.118.313816.

Kodera Y, Matsuura H, Yoshida S, Sumida T, Itakura Y, Fuwa T, Nishino H (1989) Allixin, a stress compound in garlic. *Chem Pharmacol Bull.* 37: 1656–1658.

Kojuri J, Vosoughi AR, Akrami M (2007) Effects of *Anethum graveolens* and garlic on lipid profile in hyperlipidemic patients. *Lipid Health Dis.* 6: 5. https://doi.org/10.1186/1476-512IX-6-5.

Kommula SR, Chekkilla UK, Ganugula R, Patil MA, Vadakattu SS, Myadara S, Putcha UK, Ramavat RN, Ponday LR, Tamana RR, Palla S (2020) Garlic ameliorates long-term pre diabetes induced retinal abnormalities in high fructose fed rat model. *Indian J Exp Biol.* 58(7): 452–460.

Konaklieva MI, Plotkin BJ (2006) Antimicrobial properties of organosulfur anti-infectives: A review of patent literature 1999–2005. *Recent Patentss Anti Infect Drug Discov.* 1: 177–780.

Koscielny J, Klussendorj D, Latza R, Schmitt R, Radtke H, Siegel G (1999) The anti atherosclerotic effect of *Allium sativum. Atherosclerosis.* 144: 237–249.

Kumar S (2015) Dual inhibition of acetylcholinesterase and butyrylcholinesterase enzymes by allicin. *Indian J Pharmacol.* 47(4): 444–446.

Kumar S, Mittal A, Babu D, Mittal A (2021) Herbal medicines for diabetes management and its secondary complications. *Curr Diabetes Rev.* 17(4): 437–456.

Kumar V, Kumar D (2003) *Robbins Basic Pathology.* Kumar B, Cotran RS, Stanley LR, Eds. with illustrations by Perkins JA. Saunders, Philadelphia, PA.

Kwak JS, Kim JY, Paek JE, Lee YJ, Kim HR, Park DS, Kwon O (2014) Garlic powder intake and cardiovascular risk factor: A meta-analysis of randomized controlled clinical trials. *Nutr Res Pract.* 8: 644–654. https://doi.org/10.41.62/nrp.2014.8.6.644.

Lanzotti V, Barila E, Antignani V, Bonanomi G, Scala F (2012) Antifungal saponins from bulbs of garlic, *Allium sativum.* L Var Voghiera. *Phytochem.* 78: 126–134.

Larner AJ (1995) How does garlic exert its hypocholesterolaemic action? The tellurium hypothesis. *Med Hypothesis.* 44: 295–297.

Lau BHS (2006) Suppression of LDL oxidation by garlic compounds is a possible mechanism of cardiovascular health benefit. *J Nutr.* 136(3): 7655–7685.

Lau BHS, Keeler WH, Adetumbi MH (1983) Antifungal effect of garlic. *Ann Meet Am Soc Microbiol.* 387.

Lau BHS, Tadi PP, Tosk JM (1990) *Allium satium* (garlic) and cancer prevention. *Natur Res.* 10: 937–948.

Lau BHS, Yamasaki T, Gridley DS (1991) Garlic compounds modulate macrophage and T-lymphocyte function. *Mol Biothr.* 3: 103–107.

Lawson LD (1993) Bioactive organosulfur compound of garlic and garlic products: Role in reducing blood lipids. In *Human Medicinal Agents from Plants.* Kinhorn AD, Balandrin MF, Eds. American Chemical Society, Washington, DC, pp. 306–330 (ACS Symposium Series. 534).

Lawson LD (1998) Garlic: A review of its medicinal effects and indicated active compounds. In *Phytomedicines of Europe: Chemistry and Biological Activity.* Lawson LD, Bauer R, Eds., ACS Symposium Series 691. American Chemical Society, Washington, DC, pp. 176–209.

Lawson LD (2007) Effect of consuming raw or boiled garlic on agonist-induced *ex vivo* platelet aggregation: An open trial with healthy volunteers. *FASEBJ.* 21: 864–867.

Lawson LD, Hughes BG (1992) Characterization of the formation of allicin and other thiosulfinates from garlic. *Planta Med.* 58: 345–350.

Lawson LD, Ransom DK, Hughes BG (1992) Inhibition of whole blood platelet-aggregation by compounds in garlic clove extracts and commercial garlic products. *Thromb Res.* 65: 141–156.

Lawson LD, Wood SG, Hughes BG (1991) HPLC analysis of allicin and other thiosulfinates in garlic clove homogenates. *Planta Med.* 57: 263–270.

Ledezma E, Apitz-Castro R (2006) Ajoene the main active component of garlic (*Allium sativa*): A new antifungal agent. *Rev Iberoam Micol.* 23: 75–80.

Ledezma E, Jorquera A, Bendezu H, Vivas J, Perez G (2002) Antiproliferative and leishmanicidal effect of ajoene on various *Leishmania* species: Ultra structural study. *Parasitol Res.* 88: 748–753.

Lee YL, CaEsario T, Wang Y, Shanborn E, Thrupp L (2003) Antibacterial activity of vegetables and Juices. *Nutrition.* 19: 994–996.

Lefer DJ (2007) A new gaseous signaling molecule emerges: Cardioprotective role of hydrogen sulfide. *Proc Natl Acad Sci USA.* 104: 17904–17908.

Lehmann FA (1930) Utersuchungen uber *Allium sativum* (Knoblanch). *Arch Exp Pathol Pharmacol.* 147: 245–264.

Lemar KM, Minguel AA, Sonia C, Brian O, Carsten TM, David L (2007) Diallyl disulfide depletes glutathione in *Candida albicans*: Oxidative stress mediated cell death studied by two-photon microscopy. *Yeast.* 24(8): 695–706.

Lemar KM, Turner MP, Lloyd D (2002) Garlic (*Allium sativum*) as an anti-candida agent: A fresh garlic and freeze dried extracts. *J Appl Microbiol.* 93: 398–405.

Leurcharusmee P, Sawaddirvk P, Chattipakorn N, Chattipakorn SC (2021) Possible role of garlic and its bioactive components on mitochondrial function in physiological and pathological conditions. In *Mitochondrial Physiology and Vegetal Molecules*, Oliveira MRD, Ed. Academic Press, pp. 489–539.

Li F, Zhang J, He P, Li Y (2016) Prospects of garlic sulfur compounds developing into hydrogen sulfide donor drugs. *Traditional Chin Drug Res Clin Pharmacol.* 27: 588–592. https://doi.org/10.19378/jissn.1003-9783.2016.04.025.

Li G, Qiao CH, Lin RI, Pinto J, Osborne MP, Tiwari RK (1995) Antiproliferative effects of garlic constituents in cultured human breast cancer cells. *On Col Rep.* 2: 787–791.

Li S, Guo W, Lau W, ZHang H, Zhan Z, Wang X, Wang H (2022) The association of garlic intake and cardiovascular risk factors: A systematic reviews and meta analysis. *Crit Rev Food Sci Nutr.* 63(26): 8013–8031 https://doi.1080/10408398:2022.2053657.

Li S, Yang G, Zhu X, Lin C, Sun Y, Zhao Z (2017) Combination of rapamycin and garlic-derived S-allylmercaptocysteine induces colon cancer cell apoptosis and suppresses tumor growth in Xenograft nude mice through autophagy/p62/Nrf2 pathway. *Oncol Rep.* 38: 1637–1644.

Li WQ, Zhang JY, Ma JL, Li ZX, Zhang L, Zhang Y, Guo Y, Zhou T, Li JY, Shen L, Liu WD, Han ZX, Blot WJ, Gail MH, Pan KF, Cheng W (2019) Effects of *Helicobacter pylori* supplementation on grstric cancer incidence and mortality: Folow-up of a randomized intervention trial. *BMJ.* 366: 15016. https://doi.org/10.1136/bmj.15016.

Li WR, Shi QS, Liang Q, Huang XM, Chen YB (2014) Antifungal effect and mechanism of garlic oil on *Penicillium funiculosum. Appl Microbiol Biotechnol.* 98: 8337–8346. https://doi.org/10.1007/s00253-014-5919-9.

Li X, Meng Y, Xie C, Zhu J, Wang X, Li Y, Geng S, Wu J, Zhong C, Li M (2018a) Diallyl trisulfide inhibits breast cancer stem cells via suppression of Wnt/β-catenin pathway. *Cell Biochem.* 119(5): 4134–4141. https://doi.org/10.1002/jcb.26613.

Li Z, Le W, Cui Z (2018b) A novel therapeutic anticancer property of raw garlic extract via injection but not ingestion. *Cell Death Discov.* 4: 108. https://doi.org/10.1038/s41420-018-0122-x.

Lia A, Emilia G, Ninis Y, Giftania WS (2021) *In vitro* antiplatelet activities of aqueous extract of garlic (*Allium sativum*) and black garlic in human blood. *Res J Pharm Techonol.* 15(4): 1579–1582.

Liang D, Qin Y, Zhao W, Zhai X, Guo Z, Wang R, Tong L, Lin L, Chen H, Wong Y-C, Zhong Z (2011) S-allymercapto cysteine effectively inhibits the proliferation of colorectal cancer cells under *in Vitro* and *in Vivo* conditions. *Cancer Lett.* 310(1): 69–76. https://doi.org/10.1016/j.canlet.2011.06.019.

Lindstedt S, Wlosinska M, Nilsson AC, Hllbowicz J, Fakhro M, Sheikh R (2021) Succesful improved peripheral tissue perfusion was seen in patients with atherosilerosis after 12 months of treatment with aged garlic extract. *Int Wound J.* 18(5): 681–691.

Liu C-T, Hse H, Lii C-K, Chen P-S, Sheen L-Y (2005) Effects of garlic oil and diallyl trisulfide on glycemic control in diabetic rats. *Eur J Pharmacol.* 516(2): 165–173.

Liu CT, Sheen LY, Lii CK (2007) Does garlic have a role as an antidibetic agent? *Mol Nutr Food Res.* 51: 1353–1364.

Liu C-T, Wong PL, Lii K, Hse H, Sheen LY (2006) Antidiabetic effect of garlic oil but not diallyl disulfide in rats with streptozotocin-induced diabetes. *Food Chem Toxicol.* 44(8): 1377–1384.

Liu J, Guo W, Yang ML, Liu LX, Huang SX, Tao L, Zhang F, Liu YS (2018a) Investigation of the dynamic changes in the chemical constituents of Chinese "laba" garlic during traditional processing. *Rsc Adv.* 8: 41872–41883.

Liu PZ, Nusslock R (2018) Exercise-mediated neurogenesis in the hippocampus via BDNF. *Front Neuro Sci.* 12: 52. https://doi.org/10.3389/fnins.2018.00052.

Liu SG, Ren PY, Wang GY, Yao SX, He XJ (2015) Allicin protects spinal cord neurons from glutamate-induced oxidative stress through regulating the heat shock protein 70/inducible nitric oxide synthase pathway. *Food Funct.* 6: 320–330.

Liu Y, Zhao Y, Wang Y, Zhu P, Wei Z, Wang S, Li T, Liu Z, Wu H, Sheng YL (2017) Suppressive role of diallyl trisulfide in the activated platelet-mediated hematogenous metastasis of MDA-MB-231 human breast cancer cells. *Int J Mol Med.* 39(6): 1516–1524. https://doi.org/10.3892/ijmm.2017.2953.

Liu Y, Zhao Y, Wei Z, Tao L, Sheng X, Wang S, Chen J, Ruan J, Liv Z, Cao Y, Shan Y, Wang A, Chen W, Lu Y (2018b) Targeting thioredoxin system with an organosulfur compound attenuates progression and metastasis of triple-negative breast cancer (TNBC). *Cell Physiol Biochem.* 50(5): 1945–1963. https://doi.org/10.1159/000494874.

Locatelli DA, Nazareno MA, Fusari CM, Camargo AB (2017) Cooked garlic and antioxidant activity: Correlation with organosulfur compound composition. *Food Chem.* 220: 219–224.

Loeper M, De Bray M (1921) Hypotensive effect of tincture of garlic. *Progr Med.* 36: 391–392.

Lotfi F, Abbasalipourkabir R, Goodzarzi MT, Ziamajidi N (2021) Garlic extract attenuates retinal oxidative stress and IL-1 expression in STZ-induced diabetic rats. *Med Plants Int J Phytomed Relat Ind.* 13(1): 157–163.

Lun R, Burri C, Menzinger M, Kaminsky R (1994) Antiparasitic activity of diallyltrisulfide (Dasuansu) on human and animal pathogenic protozoa (*Trypanosoma* sp., *Entamoeba histolytica* and *Giardia lamblia*) *in Vitro. Ann Soc Belg Med Tropicale.* 74: 51–59.

Luo R, Fang D, Hang H, Tang Z (2016) The mechanism in gastric cancer chemoprevention by allicin. *Anti Cancer Agents Med Chem.* 16: 802–809. https://doi.org/10.2174/187/520616666151111115443.

Lutomski J (1987) *Components and Biological Properties of Some Allium Spices.* Institute of the Medicinal Plants, Poznan, pp. 1–58.

MacDonald JA, Marchand ME, Langler RF (2004) Improving upon the *in vitro* biological activity of antithrombotic disulfide. *Blood Coagnl Fibrinolysis.* 15: 447–450.

Makheja AN, Bailey JM (1990) Antiplatelet constituents of garlic and onion. *Agents Actions.* 29: 360.

Malki A, El-saadani M, Sultan AS (2009) Garlic constituent diallyl trisulfide induced apoptosis in MCF7 human breast cancer cells. *Cancer Biol Ther.* 8(22): 2175–2185. https://doi.org/10.4161/cbt.8-22-9882.

MAlla R, Marni R, Chakraborty A, Kanal MA (2021) Dially disulfide and dially trisulfide in garlic as novel therapeutic agents to overcome drug resistance in breast cancer. *J Pharm Anal.* 12: 221–231.

Mantawy MM, Mahmoud AH (2002) Effect of *Allium cepa* and *Allium sativum* feeding on glucose, glycogen, protein bands profile and phenol oxidase activity in *Biomphalari alexandrina*. *J Egypt Soc Parasitol.* 32: 271–283.

Matsuura N, Miyamae Y, Yamnae K, Nagao Y, Hamada Y, Kawaguchi N, Katsuki T, Hirata K, Suni S, Ishikawa H (2006) Aged garlic extract inhibits angiogenesis and proliferation of colorectal carcinoma cells. *J Nutr.* 136: 8425–8465.

Maulik N (2013) *Cardiovascular Diseases: Nutritiional and Therapeutic Interventions*. CRC Press LLC, Boca Ratan, FL.

Mehrbod P, Amini E, Tavassotikheiri M (2009) Antiviral activity of garlic extract on influenza virus. *Iran J Virol.* 3: 19–23. https://doi.org/10.21859/isv.3.1.19.

Mei X, Wang MC, Xu HX, Pan XP, Gao CY, Han N, Fu MY (1982) Garlic and gastric cancer—the effect of garlic on nitrite and nitrate in gastric juice. *Acta Nutr Sinica.* 4: 53–58.

Melendez-Villanueva MA, Moran-Santibanez K, Martinez-Sanmignel JJ, Rangel-Lopez R, Garza-Navarro MA, Rodriguez-Padilla C (2019) Virucidal activity of gold nanoparticles synthesized by green chemistry using garlic extract. *Viruses.* 11: 1111. https://doi.org/10.3390/V11121111.

Miao Q, Wang R, Bai D, Xue X, Xu J, Sun X (2020) Antiartherosclerosis properties of total saponins of garlic in rats. *Evid Based Complement Altern Med.* 3683659. https://doi.org/10.1055/2020/3683659.

Miao Q, Wang R, Bai D, Xue X, Xu J, Sun X, Liu L (2020) Antiatherosclerosis properties of total saponins of garlic in rats. *Evid Based Complement Alternat Med.* 2020: 3683659. https://doi.org/10.1155/2020/3683655.

Mikaili P, Maadirad S, Moloudizargari M, Aghajanshakeri S, Sarahroodi S (2013) Therapeutic uses and pharmacological properties of garlic, shallot and their biologically active compounds. *Iran J Basic Med Sci.* 16: 1031–1048.

Milczarek M (2020) The premature senescence in breast cancer treatment strategy. *Cancers.* 12(7): 1815. https://doi.org/10.3390/cancers.120711815.

Millen AE, Subar AF, Graubard BI, Peters U, Hayes RB, Weissfeld JL, Yokochi LA, Ziegler RG (2007) Fruit and vegetable intake and prevalence of colorectal adenoma in a cancer screening trial. *Am J Clin Nutr.* 86: 1754–1764.

Milner JA (1996) Garlic: Its anticarcinogenic and antitumorigenic properties. First International conference on east-west perspectives on functional foods. *Nutr Rev.* 54: 582–586.

Milner JA (2006) Preclinical perspectives on garlic and cancer. *J Nutr.* 136: 8275–8315.

Min Li, Yun W, Wang G, Li, A, Gao J, He Q (2022) Roles and mechanisms of garlic and its extracts on atherosclerosis: A review. *Front Pharmacol.* 13: 954938. https://doi.org/10.3389/fphar.2022.954938.

Minchin WC (1912) *The Treatment, Prevention and Cure of Tuberculosis and Lupus with Allyl Sulphide*. Bailliere, Tindall and Cox, London.

Minchin WC (1927) *A Study in Tubercle Virus, Polymorphism and the Treatment of Tuberculosis and Lupus with Oleum Alli*. Bailliere, Tindall and Cox, London.

Mirelman D, Monheit D, Varon S (1987) Inhibition of growth of *Entamoeba histolytica* by allicin, the active principle of garlic extract (*Allium sativum*). *J Infect Dis.* 156: 243–244.

Mirhadi SA, Singh S (1991) Effect of garlic supplementation to cholesterol-rich diet on development of atherosclerosis in rabbits. *Indian J Exp Biol.* 29: 162–168.

Miron T, Mironchik M, Mirelman D, Wilchek M, Rabinkov A (2003) Inhibition of tumor growth by a noval approach: *In Situ* allicin generation using targeted alliinase delivery. *Mol Cancer Ther.* 2: 1295–1301.

Mizui T, Kojima M (2018) Recent advances in the biology of BDNF and the newly identified pro peptide. *J Neurol Neuromed.* 3(6): 1–4.

Mohajer ST, Ghalyanchi Langeroudi A, Karimi V, Barin A, Sadri N (2016) The effect of *Allium sativum* (garlic) extract on infectious bronchitis virus in specific pathogen free embryonic egg. *Avicenna J Phytomed.* 6: 458–467.

Mohamed N, Attia T, Ali MA (2006) Hepatoprotective activity of allicin against carbon tetrachloride induced hepatic injury in rats. *J Biol Sci.* 6: 457–468.

Mohammad SF, Brown S, Chuang HYK, Mason RG (1980) Isolation, characterization, identification and synthesis of an inhibitor of platelet function from *Allium sativum. Fed Proc.* 39: 543A.

Mohammad SF, Woodward SC (1986) Characterization of a potent inhibitor of platelet aggregation and release reaction isolated from *Allium sativum* (garlic). *Thromb Res.* 44: 793–796.

Mohana L, Sandhya R, Kiran UA (2012) A review on diabetes milletus and the herbal plants used for its treatment. *Asian J Pharm Clin Res.* 5(4): 15–21.

Moosavian SP, Arab A, Paknahad Z, Moradi S (2020) The effects of garlic supplementation on oxidative stress markers: A systematic review and metaanalysis of randomized controlled trials. *Complement Ther Med.* 50: 102385. https://doi.org/10.1016/j.ctim.2020.102385.

Morbidone L, Arterburn JM, Young V, Mallins D, Mulrow C, Lawrence V (2001) Garlic: Its history and adverse effects. *J Herb Pharmacother.* 1: 63–83.

Moriguchi T, Matsuura H, Itakura Y, Katsuki H, Saito H, Nishiyama N (1997) Allixin, a phytoalexin produced by garlic and its analogues as novel exogenous substances with neurotrophic activity. *Life Sci.* 61(14): 1413–1420.

Mukherjee D, Banerjee S (2013) Learning and memory promoting effects of crude garlic extract. *Indian J Exp Biol.* 51: 1094–1100.

Mukherjee I, Lekli I, Goswami S, Das DK (2009) Freshly crushed garlic is a superior cardioprotective agent than processed garlic. *J Agric Food Chem.* 57: 7137–7144.

Murugavel P, Pari L, Sitasawad SL, Kumar S, (2007) Cadmium induced mitochondrial injury and apoptosis in vero cells: Protective effect of diallyl tetrasulfide from garlic. *Int J Biochem Cell Biol.* 39: 161–170.

Myneni AA, Shen-chih C, Rungui N, Li L, Mya KS, Jiawei Li, Jia SU, Gary AG, Shunzhang Yu, Zuo-Feng Z, Mu L (2016) Raw garlic consumption and lung cancer in a Chinese population. *Cancer Epidemiol Biomarkers Prev.* 25(4): 624–633. https://doi.org/10.1158/1055-9965.EPI-15-0760.

Na HK, Kim EH, Choi MA, Park JM, Kim DH, Surh YJ (2012) Diallyl trisulfide induces apoptosis in human breast cancer cells through ROS-medicated activation of JNK and AP–1. *Biochem Pharmacol.* 80(10): 1241–1250. https://doi.org/10.1016/J.bcp.2012.08.024.

Nagai K (1973) Experimental studies on the preventive effect of garlic extract against infection with influenza virus. *JPN J Infect Dis.* 47: 321–325.

Nagini S (2008) Cancer chemoprevention by garlic and its organosulfur compound—panacea or promise? Anticancer Agents. *Med Chem.* 8: 313–321.

Naji KM, Shukri E, Shaibani AI, Fatima AA, Safa'a AA-S, D'souza MR (2017) Hepatoprotective and antioxidant effects of single clove garlic against CCl4-induced hepatic damage in rabbits. *BMC Compound Alter Med.* 17: 411. http://doi.org/10.1186/S/2906-017-1916-8.

Ngo SN, Williams DB, Cobiac L, Head RJ (2007) Does garlic reduce risk of colorectal cancer? A systematic review. *J Nutr.* 137: 2264–2269.

Nies LK, Cymbals AA, Kasten SL, Lamprecht DG, Olson KL (2006) Complementary and alternative therapies for the management of dyslipidemia. *Ann Pharmacother.* 40: 1984–1992.

Nishida M, Hada T, Kuramochi K, Yoshida H, Yonezawa Y, Kuriyama I, Sugawara F, Yoshida H, Mizushina Y (2008) Diallyl Sulfides: Selective inhibitors of family X DNA polymerases from garlic (*Allium sativum* L). *Food Chem.* 108: 551–560.

Nkrumah-Eli YM, Reuben JS, Hudson A, Taka E, Badisa R, Ardley T, Israel B, Sadrud-Din SY, Oriaku E, Darling-Reed SF (2012) Diallyl trisulfide as an inhibitor of benzo pyrene-induced precancerous carcinogenesis in MCF-10A cells. *Food Chem Toxicol.* 50(7): 2524–2530. http://doi.org/10.1016/j.fct.2012.04.010.

O'Gara EA, Hill DJ, Maslin DJ (2000) Activitis of garlic oil, garlic powder and their diallyl constituents against *Helicobacter pylori. Appl Environ Microbiol.* 66: 2269–2273.

O'Gara EA, Maslin DJ, Nevill AM, Hill DJ (2008) The effect of simulated gastric environments on the anti-Helicobacter activity of garlic oil. *J Appl Microbiol.* 104: 1324–1331.

Ogita A, Fugita K, Taniguchi M, Tanaka T (2006) Dependence of synergistic fungicidal activity of Cu2+ and allicin, an allyl sulfur compound from garlic, on selective accumulation of the ion in the plasma membrane fraction via allicin-mediated phospholipid peroxidation. *Planta Med.* 72: 875–880.

Ohta R, Yamada N, Kaneko H, Ishikawa K, Fukuda H, Fujino T, Suzuki A (1999) *In vitro* inhibition of the growth of *Helicobacter pylori* by oil-macerated garlic constituents. *Antimicrob Agents Chemother.* 83: 1811–1812.

Ojo OA, Adegboyega AE, Johnson GI, Umedum NL, Onuh K, Adeduro MN, Nwobodo UO, Elekan AO, Alemika TE, Johnson TO (2021) Deciphering the interactions of compounds from *Allium sativum* targeted towards identification of novel PTP IB inhibitors in diabetes treatment: A computational approach. *Inform Med Unlocked.* 26: 100719. http://doi.org/10.1016/j.imy.2021.100719.

Okue S, Yaguchi M, Miura A, Ozaki-Masuzawa Y, Hosono T, Seki T (2022) The garlic—derived organosulfur compound diallyl trisulphide suppresses tissue factor function. *Food Funct.* 13: 1246–1255. https://doi.org/10.1039/d1f00226g.

Omar SH (2013) Garlic and cardiovascular diseases. *Nat Prod Springer.* 3661–3696.

Omar SH, Hasan A, Hunjul N, Ali J, Aqil M (2007) Historical, chemical and cardiovascular perspectives on garlic: A review. *Pharmacog Rev.* 1: 80–87.

Oommen S, Anto RJ, Srinivas G, Karungaran D (2004) Allicin (from garlic) induced caspace-mediated apoptosis in cancer cells. *Eur J Pharmacol.* 485: 97–130.

Orekhov AN, Tertov VV, Sobenin IA, Pivovarova EM (1995) Direct anti-atherosclerosis-related effects of garlic. *Ann Med.* 27: 63–65.

Padiya R, Khatua TN, Bagul PK, Kuncha M, Banerjee SK (2011) Garlic improves insulin sensitivity and associated metabolic syndromes in fructose fed rats. *Nutr Metab.* 8(1): 1–8.

Pandey P, Khan F, Alshammari N, Saeed A, Aqil F, Saeed M (2023) Updates on the anticancer potential of garlic organosulfur compounds and their nanoformulations: Plant therapeutics in cancer management. *Front Pharmacol.* http://doi.org/10.3389/fphar.2023.1154034.

Pandey P, Khan F, Kumar A, Srivastava A, Jha NK (2021) Screening of potent inhibitors against 2019 novel corona virus (Covid-19) from *Allium sativum* and *Allium cepa*: An in silico approach. *Biointerface Res Appl Chem.* 11: 7981–7993. https://doi.org/10.33263/BRIAC111.79817993.

Pandiya R, Banerjee SK (2013) Garlic as an anti-diabetic agent: Recent progress and patent review. *Recent Pat Food Nutr Agric.* 5(2): 105–127.

Papayannopoulos G (1969) Garlic. *Lancet.* 2: 962.

Parham M, Bagherzadeh M, Asghari M, Akbari H, Hosseini Z, Rafiee M, Vafaimanesh J (2020) Evaluating the effect of a herb on the control of blood glucose and insulin-resistance in patients with advanced type 2 diabetes (a double-blind clinical trial). *Caspian J Intern Med.* 11(1): 12.

Park BM, Chun H, Chae SW, Kim SH (2017) Fermented garlic extract ameliorates monocrotline-induced pulmonary hypertension in rats. *J Funct Foods.* 30: 247–253.

Pâruv M, Mot CA, Parvu AE, Mircea C, Socber L, Rosca-Casian O, Tigu AB (2019) *Allium sativum* extract chemical composition, antioxidation activity and antifungal effect against *Meyerozyma guilliermondii* and *Rhodotorula mucilaginosa* causing onychomysis. *Molecules.* 24: 3958. https://doi.org/10.3390/molecules24213958.

Pasteur L (1858) Memorie sur la fermentation appelee lactique. *Ann Chim Phys.* 52: 404–418.

Patya M, Zahalka MA, Vanichkin A, Rabinkov A, Miron T, Mirelman D, Wilchek M, Lander HM, Novogrodsky A (2004) Allicin stimulates lymphocytes and elicites an antitumor effect: A possible role of p21 ras. *Int J Immunol.* 16: 275–281.

Pectov U (1979) Plants with hypotensive antiatheromatous and coronarodilating action. *Am J Chin Med.* 7: 197–236.

Perchellet JP Perchellet EM, Abney NL, Zirnstein JA, Belman S (1986) Effects of garlic and onion oils on glutathione peroxidase activity, the ratio of reduced/oxidized glutathione and ornithine decarboxylase induction in isolated mouse epidermal cells treated with tumor promotors. *Cancer Biochem Biophys.* 8(4): 299–312.

Perkin DM, Laora E, Muir CS (1988) Estimates of worldwide frequency of sixteen major cancers in 1980. *Int J Cancer.* 41: 184–197.

Petropoulos S, Fernandes A, Barros L, Ciric A, Sokovic M, Ferreira I (2018) Antimicrobial and antioxidant properties of various Greek garlic genotypes. *Food Chem.* 245: 7–12.

Petrovic V, Nepal A, Olaisen C, Bachke S, Hira J, Sogaard CK, Rost LM, Misund K, Andreassen T, Melo TM, Bartsova Z, Bruheim P, Otterlei M (2018) Anti-cancer potential of homemade fresh garlic extract is related to increased endoplasmic reticulum stress. *Nutruents.* 10: 450. https://doi.org/10.3390/nu10040450.

Phan ADT, Netzel G, Chhim P, Netzel ME, Sultanbawa Y (2019) Phytochemical characteristics and antimicrobial activit of Australian grown garlic (*Allium sativum*) cultivars. *Foods.* 8: 358. https://doi.org/10.3390/foods8090358.

Piotrowski G (1948) L'ail on therapeutique. *Praxix.* 26: 488–492.

Piscitelli SC, Burstein AH, Welden H, Gallicano KD, Falloon J (2002) The effect of garlic supplement on the pharmaco kinetics of saquinavir. *Clin Infect Dis.* 34: 234–238.

Polat AZ, Vural A, Ozan F, Tepe B, Oezcelik S, Cetin A (2008) *In vitro* evaluation of the amoebicidal activity of garlic (*Allium sativum*) exract on *Acanthamoeba castellanii* and its cytotoxic potential on corneal cell. *J Ocular Pharmacal Therapeut.* 24: 8–14.

Powolny AA, Singh SV (2008) Multitargeted prevention and therapy of cancer by diallyl trisulfide and related *Allium* vegetable—derived organosulfur compounds. *Cancer Lett.* 269: 305–314.

Powolny AA, Singh SV (2008) Multitargeted prevention and therapy of cancer by diallyl trisulfide and related *Allium* vegetable-derived organosulfur compounds. *Cancer Lett.* 269: 305–314.

Prager-Khoutorsky M, Gonharov I, Rabinko A, Mirelman D, Geiger B, Bershadsky AD (2007) Allicin inhibits cell polarization, migration and division *via* its direct effect on microtubules. *Cell Motil Cytoskeleton.* 64: 321–327.

Prasad K, Laxdal VA, Yu M, Raney BL (1995) Antioxidant activity of allicin, an active principle in garlic. *Mol Cell Biochem.* 148: 183–189. http://doi.org/10.1007/BF00928155.

Qi R, Liao F, Inoue K, Yatomi Y, Sato K, Ozaki Y (2000) Inhibition by diallyl trisulfide, a garlic component of intracellular Ca2+ mobilization without affecting inositol-1,4,5-triphosphate (IP3) formation in activated platelets. *Biochem Pharmacol.* 60: 1475–1483.

Qidwai W, Ashfaq T (2013) Role of garlic usage in cardiovascular diseasese prevention: An evidence based approach. *Evid Based Complement Altern Med.* 125649, 1–9.

Qu Z, Mossine VV, Cui J, Sun GY, Gu Z (2016) Protective effects of AGE and its components on neuroinflammation and neurodegeneration. *Nuromolecular Med.* 18(3): 474–482.

Rahman K, Lowe GM (2006) Garlic and cardiovascular disease: A critical review. *J Nutr.* 136: 7365–7405.

Reda R, Alshiman A Khalil, Mohamed E, Safaa IT, Enas AR (2024) Anti-parasitic activity of garlic (*Allium sativum*) and onion (*Allium cepa*) extracts against Dactylogyrus Spp. (*Monogenean*) in Nile tilapia (*Oreochromis niloticus*): Hematology, immune response, histopathological investigation and inflammatory cytokine genes of gills. *BMC Vet Res.* 20: 334. https://doi.org/10.1186/s/2917-024-04187-5.

Reinhart KM, Coleman CI, Teevan C, Vachhani P, White CM (2008) Effects of garlic on blood pressure in patients with and without systolic hypertension: A meta-analysis. *Ann Pharmacother.* 42: 1766–1771.

Reuter HD, Deninger R, Wagner H, Linker H (1974) Investigations on the platelet aggregation inhibiting principle of garlic extract. In *Report of the 15th International Congress of the Society of Hematology.* Jerusalam.

Reuter HD, Koch HP, Lawson LD (1996) Therapeutic effect and application of garlic and its preperations. In *The Science and Therapeutic Application of Allium Sativum L. and Related Spices.* Koch HP, Lawson LD, Eds. Williams and Wilkins, Baltimore, MD, pp. 135–212.

Reuter HD, Sendl A (1995). *Allium sativum* and *Allium ursinum*: Part 2. Pharmacology and medicinal application. *Phytomedicine.* 2: 73–91.

Ribeiro M, Alvarenga L, Cardozo LF, Chermut TR, Sequeira J, Moreira LDSG, Teixeira KTR, Shiels PG, Stenvinkel P, Mafra D (2021) From the distinctive smell to therapeutic effects: Garlic for cardiovascular, hepatic, gut, diabetes and chronic kidney disease. *Clin Nutr.* 40(7): 4807–4819.

Ried K (2016) Garlic lowers blood pressure in hypertensive individuals, regulates serum cholesterol and stimulates immunity: An updated meta-analysis. *Nutricion.* 146: 3895–3965.

Ried K, Fakler P (2014) Potential of garlic (*Allium sativum*) in lowering high blood pressure: Mechanisms of actions and clinical relevance. *Inter Blood Press Control.* 7: 71–82.

Ried K, Frank OR, Stocks NP (2013a) Aged garlic extract reduce blood pressure in hypertensives: A dose-response trial. *Eur J Clin Nutr.* 67: 64–70.

Ried K, Frank OR, Stocks NP, Fakler P, Sullivan T (2008) Effect of garlic on blood pressure: A systematic review and meta-analysis. *BMC Cardiovasc Disord.* 8: 1. http://www.biomedcentral.com/1471-2261/8/13.

Ried K, Toben C, Falker P (2013b) Effect of garlic on serum lipids: An update meta-analysis. *Nut Rev.* 71: 282–299.

Rosas-Gonzalez VC, Tellez-Banuelos MC, Hernandez-flores G, Bravo-Cuellar A, Aguilar-Lemarroy A, Jave-Suarez LF Haromati J, Ibarra FS, Ortiz-Lazareno PC (2020) Differential effects of alliin on apoptosis and senescence in luminal A and triple-negative breast cancer: Caspase and pro-apoptotic gene involvement. *Fundam Clin Pharmacol.* 34(6): 671–686. http://doi.org/10.1111/fcp.12559.

Ross IA (1999) *Medicinal Plants of the World Chemical Constituents Traditional and Modern Medicinal Uses.* Human Press, Totowa.

Rouf R, Uddin SJ, Sarkar DK, Islam MT, Ali ES, Shilpi JA (2020) Antiviral potential of garlic (*Allium sativum*) and its organosulfur compounds: A systematic update of pre-clinical and clinical data. *Trends Food Sci Technol.* 104: 219–234. http://doi.org/10.1016/j.tifs.2020.08.006.

Ryu JH, Kang D (2017) Physiochemical properties, biological activity, health benefits and general limitations of aged black garlic: A review. *Molecules.* 22(6): 919.

Sabitha P, Adhikari PM, Shenoy SM (2005) Efficacy of garlic paste in oral candidiasis. *Trop Doct.* 35: 99–100.

Saikat ASM, Hossain R, Mina FB, Das S, Khan IN, Mohammad SM, Muhamad TI (2022) Antidiabetic effect of garlic. *Rev Bras Farmacogn.* 32(1): 11. https://doi.org/10.1007/s43450-021-00193-Y.

Sainani GS, Desai DB, Gorhe NH, Natu SM, Pise DV, Sainani PG (1979) Effect of dietary garlic and onion on serum lipid profile in Jain community. *Indian J Med Res.* 69: 776–780.

Sanie-Jahromi F, Zia Z, Afarid M (2023) A review on the effect of garlic on diabetes BDNF and VEGF as a potential treatment for diabetic retinopathy. *Chin Med.* 18(18): 2023. http://doi.org/10.1186/s13020-023-00725-9.

Sasaki J, Kita J (2003) Bacteriocidal activity of garlic powder against *Bacillus anthracis. J Nutr Vitaminol.* 49: 297–299.

Sawai T, Itoh Y, Ozaki H, Isoda N, Okamoto K, Kashima Y, Takeuchi Y, Kida H, Ogasawara K (2008) Induction of cytotoxic T-lymphocyte and antibody responses against highly pathogenic avian influenza virus infection in mice by inoculation of a pathogenic H5N1 influenza virus particles inactivated with formalin. *Immunology.* 124: 155–165.

Saxena KK, Gupta B, Kulshreshtha VK, Srivastava RK, Prasad DN (1980) Effect of garlic treatment on isoprenaline—induced myocardial necrosis in Albion rats. *Indian J Physiol Pharmacol.* 24: 233–236.

Schaffer EM, Liu JZ, Green J, Dangler CA, Milner JA (1996) Garlic and associated allyl sulfur conponents inhibit N-methyl-N-nitrosourea induced rat mammary carcinogensis. *Cancer Lett.* 102(1–2): 199–204. https://doi.org/10.1016/0304-3835(96)04160-2.

Schmitt B, Bernhardt T, Moellar HJ, Heuser I, Frolich L (2004) Combination therapy in Alzheimer's. *CNS Drugs.* 18: 827–844.

Semenkovich K, Brown ME, Sarakic DM, Lustman PJ (2015) Depression in type 2 deabetes mellitus: Prevalence impact and treatment. *Drugs.* 75(6): 577–587.

Semuyaba I, Safiriyu AA, Tiyo EA, Niurka RF (2017) Memory improvement effect of ethanol garlic (*A. sativum*) extract in streptozotocin-nicotinamide induced diabetic wistar rats is mediated through increasing hippocampal sodium-potassium ATPase, glutamine synthetase and calcium ATPase activities. *Evid-based Complement Altern Med.* 2017: 1–7. Article ID 3720380. https://doi.org/10.1155/2017.3720380.

Sengupta A, Ghosh S, Bhattacharjee S (2004) *Allium* vegetables in cancer prevention: An overview. *Asion Pac J Cancer Prev.* 5: 237–245.

Shabani E, Sayemiri K, Mohammmadpour M (2018) The effect of garlic and lipid profile and glucose parameters in diabetic patients: A systematic review and meta analysis. *Prim Care Diab.* http://doi.org/10.1016/j.pcd.2018.07.007.

Shadkchan Y, Shemesh E, Mirelman D, Miron T, Rabinkov A, Wilchek M, Osherov N (2004) Effecacy of allicin, the reactive molecule of garlic, inhibiting *Aspergillus* spp. *in vitro*, and in a murine model of disseminated aspergillosis *J Antimicrob Chemother.* 53: 832–836.

Shafeeque K, Hashim KM (2018) Comparative anti-angiogenesis study between allicin nanoparticle and normal allicin from garlic (*Allium sativum* Linn.). *Eur J Exp Biol.* 8(4): 27. http://doi.org/10.21767/2248.9215.100068.

Shahzad M, Iqbal H, Akbar M, Ahmad H, Huma S, Qdir M, Khan MZ, Akbar A, Khan RMI, Jadoon SK (2023) Effect of garlic consumption on hyperlipidemia patients: Observational cohort study. *J Popul Ther Clin Pharmacol.* 30(17): 1635–1643.

Shaker M, Khamisipour G, Sadeghipour H, Zar A, Naeimi B, Akbarzadeh S (2022) Effect of resistance training and garlic extract on insulin sensitivity/resistance and biochemical parameters in diabetic rats. *Comp Exerc Physiol.* 18(2): 163–170.

Sham-Ghahfarokhi M, Shokoohamiri MR, Amirrajab N, Moghadasi B, GhAjari A, Zeini F (2006) *In vivo* antifungal activities of *Allium cepa*, *Allium sativum* and keto canazole against some pathogenic yeast and dermatophytes. *Fitoterapia.* 77(4): 321–323.

Shang A, Cao SY, Xu XY, Gan RY, Tang GY, Corke H, Mavumengwana V, Hua-BLi (2019) Bioactive compounds and biological functions of garlic (*Allium sativum* L.). *Foods.* 8(7): 246. https://doi.org/10.3390/foods8070246.

Sharma N (2019) Efficasy of garlic and onion against viruses. *Int J Res Pharmaceutical Sci.* 10: 3578–3586. https://doi.org/10.26452/ijrps.v1oi4.1738.

Sharma VC, Sethi MS, Kumar A, Rarotra JR (1977) Antibacterial property of *Allium sativum* L.: *In vivo* and *in vitro* studies. *Indian J Exp Biol.* 15: 466–468.

Shaw JE, Sicree RA, Zimmet PZ (2010) Global estimate of the prevalence of diabetes for 2010 and 2030. *Deabetes Res Clin Pract.* 87(1): 4–14.

She J, Yang, P, Hong R, Bai C (2013) Lung cancer in China: Challenges and interventions. *Chest.* 143: 1117–1126.

Sheela C, Kumud K, Augusti K (1995) Anti-diabetic effects of onion and garlic sulfoxide amino acids in rats. *Planta Med.* 61(4): 356–357.

Shoji S, Furuishi K, Yanase R, Miyazaka T, Kino M (1993) Allyl compounds selectively killed human immunodeficiency virus (Type-1)—infected cells. *Biochem Biophys Res Commun.* 194: 620–621.

Shokrzadeh M, Ebadi AG (2006) Antibacterial effect of garlic (*Allium sativum* L.) on staphylococcus aureus. *Pak J Biol Sci.* 9: 1577–1579.

Shukla Y, Kalra N (2007) Cancer chemoprevention with garlic and its constituents. *Cancer Lett.* 247: 167–181.

Si L, Lin R, Jia Y, Jian W, Yu Q, Wang M, Yang S (2019) Lactobacillus bulgaricus improves antioxidant capacity of black garlic in the prevention of gestational diabetes mellitus: A randomized control trial. *Biosci Rep.* 39(8): BSR20182254. https://doi.org./10.1042/BSR20182254.

Siegel G, Malmsten M, Pietzsch J, Schmidt A, Buddecke E, Michel F, Plock M, Schneider W (2004) The effect of garlic on arteriosclerotic nanoplaque formation and size. *J Phytomed.* 11: 24–35.

Siess MH, Le Bon AM, Teyssier C, Belloir C, Singh V, Berges R (2007) Garlic and cancer. *Med Aromatic Plant Sci Biotech.* 1: 25–30.

Silprasit K, Seetaha S, Pongsanarakul P, Hannongbua S, Choowongkomon K (2011) Anti-HIV-1 reverse transcriptase activities of hexane extracts from some Asian medicinal plants. *JMPR.* 5: 4899–4906.

Singh DK, Porter TD (2006) Inhibition of sterol 4α methyl oxidase is the principal mechanism by which garlic decreases cholesterol synthesis. *J Nutr.* 136: 7595–7645.

Singh RN, Kumar P, Kumar N, Singh DK (2020) Garlic (*Allium sativum*): Pharmaceutical uses for human health. *Int J of Pharmaceutical Sci Res.* 11(9): 4214–4228.

Singh V, Belloir C, Siess MH, Le Bon AM (2006) Inhibition of carcinogen-induced DNA damage in rat liver and colon by garlic powders with varying alliin content. *Nutr Cancer.* 55: 178–184.

Singh VK, Singh DK (1996b) Enzyme inhibition by allicin, the molluscicidal agents of *Allium sativum* L. (Garlic). *Phytother Res.* 10: 383–386.

Singh VK, Singh DK (2008) Pharmacological effects of garlic (*Allium sativum* L.). *ARBS Ann Rev Biomed Sci.* 10: 6–26.

Sivam GP (2001) Protection against *Helicobacter pylori* and other bacterial infections by garlic. *J Nutr.* 131: 11065–11085.

Sivam GP, Lampe JW, Ulness B, Swanzy SR, Potter JD (1997) *Helicobacter pylori—in vitro* susceptibility to garlic (*Allium sativum*) extract. *Nutr Cancer.* 27: 118–121.

Sivasami K (2017) Scientometric analysis of garlic research publications. *Glob Res Anal.* 6(6): 194–197. ISSN No. 2277–8180.

Sobenin I, Andrianova I, Ionova V, Karagodin V, Orekhov A (2012) Anti-aggregatory and fibrinolytic effects of time-released garlic powder tablets. *Med Health Sci J.* 10: 47–51.

Sobenin IA, Nedosygova LV, Filatova LV, Balabolkin MI, Gorchakova TV, Orekhov AN (2008) Metabolic effects of time-released garlic powder tablets in type 2 diabetes mellitus: The results of double-blinded placebo-controlled study. *Acta Diabetol.* 45: 1–6.

Sobenin IA, Pryanishnikov VV, Kunnova LM, Rabinovich YA, Martirosyan DM, Orekhov AN (2010) The effect of time-released garlic powder tablets on multifunctional cardiovascular risk in patients with coronary artery disease. *Lipid Health Dis.* 9: 119.

Soffar SA, Mokhtar GM (1991) Evaluation of the antiparasitic effect of aqueous garlic (*Allium sativum*) extract in hymenolepiasis nana and giardiasis. *J Egypt Soc Parasitol.* 21(2): 497–508.

Song H, Cui J, Mossine VV, Greenlief CM, Fritsche K, Sun GY, Zezong GU (2020) Bioactive components from galic on brain resiliency against neuroinflammation and neurodegenration. *Exp Ther Med.* 19(2): 1554–1559.

Sparnins VL, Barany G, Wattenberg LW (1988) Effects of organosulfur compounds from garlic and onion on benzo(a)pyrene-induced neoplasia and glutathione S-transferase activity in the mouse. *Carcinogenesis.* 9: 134.

Sparnins VL, Mott AW, Watenberg AW (1986) Effects of allylmethyl trisulfide on glutathione S-transferase activity and PB-induced neoplasia in the mouse. *Nutr Cancer.* 8: 211–215.

Srinivasan VA (1969) A new antihypertensive agent. *Lancet.* 2: 800.

Srivastava KC (1984) Effect of aquous extracts of onion, garlic and ginger on platelet aggregation and metabolism of arachidonic acid in blood vascular system: *In Vitro* study. *Prostaglandin Leukot Med.* 13: 227–235.

Srivastava KC, Bordia A, Verma SK (1995) Garlic (*Allium sativum*) for disease prevention. *South Afr J Sci.* 91: 68–77.

Srivastava KC, Justesen V (1989) Isolation and effect of some garlic components on platelet aggreagation and metabolism of arachidonic acid in human blood platelets. *Wien Klin Wochenschr.* 8: 293–299.

Srivastava KC, Mustafa T (1993) Pharmacological effects of spices: Eicosanoid modulating activities and their significance in human health. *Biomed Rev.* 2: 15–19.

Srivastava KC, Tyagi OD (1993) Effect of garlic derived principle (ajoene) on aggregation and arachidonic acid metabolism in human blood platelets. *Prostaglandins Leukot Essent Fatty Acids.* 49: 587–595.

Stabler SN, Tejani AM, Huynh F, Fowkes C (2012) Garlic for the prevention of cardiovascular morbidity and mortality in hypertensive patients. *Cochrane Database Syst Rev.* 15(8). https://doi.org/10.1002/14651858.CD007653.pub2.

Stan SD, Kar S, Stoner GD, Singh SV (2008) Bioactive food components and cancer risk reduction. *J Cell Biochem.* 104: 339–356.

Stephen RD (2005) An overview of the antifungal properties of allicin and its breakdown products: The possibility of a safe and effective antifungal prophylactic. *Mycoses.* 48: 95–100.

Stettler C, Allemann S, Juni P, Cull CA, Holman RR, Egger M, Krahenbuhl S, Diem P (2006) Glycemic control and macrovascular disease in types 1 and 2 diabetes mellitus: Meta-analysis of randomized trials. *Am Heart J.* 152(1): 27–38.

Stolooff L (1976) Occurrence of mycotoxins in food and feeds. In *Mycotoxins and Other Fungal Related Food Problems.* Rodricks JJ, Ed. American Chemical Society, Washington, DC, pp. 23–50.

Sujithra K, Srinivasan S, Indumathi D, Kumar V (2018). Allyl methyl sulfide, an organosulfur compound alleviates hyperglycemia mediated hepatic oxidative stress and inflammation in streptozotocin-induced experimenta rats. *Biomed Pharmacother.* 107: 292–302.

Sumioka J, Hayama M, Shimokawa Y, Shiraishi S, Tokunaga A (2006) Lipid—lowering effect of monascus garlic fermented extract (MGFE) in hyperlipidemia subjects. *Hiroshima J Med Sci.* 55: 59–64.

Sumiyoshi H, Wargovich MA (1989) Garlic (*Allium sativum*): A review of its relationship to cancer. *Asia Pac J Pharmacol.* 4: 133–140.

Swanston-Flatt S, Day C, Baiely C, Flatt P (1990) Traditional plant treatments for diabetes. Studies in normal and streptozotocin diabetic mice. *Diabetologia.* 33(8): 462–464.

Tadi PP, Lav BHS, Teel RW, Herrmann CE (1991a) Binding of Aflagoxin B1 to DNA inhibited by ajoene and diallyl sulfide. *Anti Cancer Res.* 11: 2037–2042.

Tadi PP, Lav BHS, Teel RW, Herrmann CE (1991b) Organosulfur compound of garlic modulate mutagenesis, metabolism, and DNA binding of Aflatoxin B1. *Nutr Cancer.* 15: 87–95.

Tadi PP, Robert MS, Teel W, Lau BHS (1990) Anticandidal and anticarcinogenic potential of garlic. *Int Clin Nutr Rev.* 10: 223–229.

Tadu CK, Dutta T, Ghorai M, Biswas P, Samanta D, Oleksak P, Jha NK, KuMar M, Radha, Prockow J, Perez dela Lastra JM, Dey A (2022) Traditional uses, Phytochemistry, Pharmacology and Toxicology of garlic (*Allium sativum*), a store house of diverse phytochemicals. A review of research from the last decade focusing on health and nutritional implications. *Front Nutr.* 1–19. https://doi.org/10.3389/fnut2022.929554.

Tahiliani P, Kar A (2003) The conditional effects of *Triginella* and *Allium* extracts in the regulation of hyper thyroidism in rats. *Phytomedicine.* 10: 665–668.

Takashima M, Kanamori Y, Kodera Y, Morihara N, Tamura K (2017) Aged garlic extract exerts endotherium-dependent vasorelaxant effect on rat aorta by increasing nitric oxide production. *Phytomedicine.* 24: 56–59 https://doi.org/10.1016/J.phymed.2016.11.016.

Takim K, Yigin A, Koyuncu I, Kaya R, Gulcin I (2021) Anticancer, Anticholinesterase and anti diabetic activities of tunceli garlic (*Allium tuncelianum*): Determining its phytochemical content by LC-MS/MS analysis. *J Food Meas Charact.* 15(4): 3325–3335.

Tasai Y, Col LL, Davis LF, Lockwood SJ, Simmons V, Wild GC (1985) Antiviral properties of garlic: *In vitro* effects on influenza B., herpes simplex and coxsackie viruses. *Planta Med.* 8: 460–461.

Tatarintsev AV, Vrzhets PV, Ershow DE, Schegalev AA, Turgiev A, Karamov EV (1992) The ajoene blockade of integrin-dependent processes in an HIV-infected cell system. *Vestn Ross Aked Med Nauk.* 11(12): 6–10.

Tedeschi P, Nigro M, TrAvagli A, Catani M, CAvazzini A, Merighi S, Stefania G (2022) Therapeutic potential of allicin and aged garlic extract in Alzheimer's disease. *Int J Mol Sci.* 23(13): 6950. https://doi.org/10.3390/ijms.3136950.

Tellez TME, Aguilar ESV, Rojas MTA, Guineo DOD, Quevedo R, Diaz O, Montes JMB (2020) Garlic (*Allium sativum* L.) and its beneficial properties for health: A review. *Agroind Sci.* 10(1): 103–115.

Teranishi R, Apitz-Castro R, Robson SC, Ramano E, Cooper DKC (2003) Inhibition of baboon platelet aggregation *in Vitro* and *in Vivo* by the garlic derivative, ajoene. *Xenotransplantation.* 10: 374–379.

Tesfaye A (2021) Revealing the therapeutic uses of garlic (*Allium sativum*) and its potential for drug discovery. *Sci World J.* 2021, article ID 8817288: 7. https://doi.org/10.1155/2021/8817288.

Tesfaye A, Mengesha W (2015) Traditional uses, phytochemistry and pharmacological properties of garlic (*Allium sativum*) and its biological active compounds. *Inter J Sci Res Sci Eng Technol.* 5: 142–148.

Thomson M, Al-Amin ZM, Al-Qattan KK, Shaban LH, Ali M (2007) Antidiabetic and hypolipidaemic properties of garlic (*Allium sativum*) in streptozotocin-induced diabetic rats. *Int J Diabetes Metab.* 15(3): 108–115.

Thomson M, Ali M (2003) Garlic (*Allium sativum*): A review of its potential use as an anticancer agent. *Curr Cancer Drug Targets.* 3: 67–81.

Tilli CMLJ, Stavast-Kooy AJ, Vuerstaek JDD, Thissen MRTM, Krekels GAM, Ramaekers FCS, Neumann HAM (2003) The garlic derived organosulfur component ajoene decreases basal cell carcinoma tumor size by inducing apoptosis. *Arch Dermatol Res.* 295: 117–123.

Torre IA, Bray F, Siegel RL, Ferlay J, Lortet-Tieulent J, Jemal A (2015) Global cancer statistics. *CA Cancer J Clin.* 65: 87–108.

Torres E, Grimaldo Q, Henry J, Carlos R, Heyul C, Francisco D (2024) Phytochemical profile and antimicrobial activity of aqueous garlic extract (*Allium sativum* L.) for food preservation. *Int J Eng Trends Technol.* 72(8): 8–17.

Tsai C-Y, Wen S-Y, Shibu M-A, Yang Y-C, Peng H, Wang B, Wei Y-M, Chang HYu, Lee C-Yu, Huang C-Y, Kyo WW (2015) Diallyl trisulphide protects against high glucose-induced cardiac apoptosis by stimulating the production of cystathionine gamma-lyase derived hydrogen sulfide. *Int J Cardiol.* 195: 300–310.

Tsai TH, Tsai PJ, Ho HC (2005) Anti-oxidant and anti-inflammatory activities of several commonly used spices. *J Food Sci.* 70(1): 93–97. http://doi.org/10.1111/j1365-2621.2005tb09028x.

Tsao SM, Hsu CC, Yin MC (2003) Garlic extract and two diallyl sulphides inhibits methicillin-resistant *Staphylococcus aurens* infection in BALB/A mice. *J Antimicrob Chemother.* 52: 974–980.

Tsubura A, Lai YC, Kuwata M, Uehara N, Yoshizawa K (2011) Anticancer compounds for breast cancer control. *Anti-cancer Agents Med Chem.* 11(3): 249–253. https://doi.org/10.2174/187152011795347441.

Tutakne MA, Satyanarayanan G, Bharadwaj JR, Sethi IC (1983) Sporotrichosis treated with garlic juice: A case report. *Indian J Dermatol.* 28: 41–45.

Urbán N, Guillemot F (2014) Neurogenesis in the embryonic and adult brain: Same regulators, different roles. *Front Cell Neuro Sci.* 8: 396. https://doi.org/10.3389/fncel.2014.00396.

Varshney R, Budoff MJ (2016) Garlic and heart disease. *J Nutr.* 146: 4165–4215.

Vijayakumar S, Malaikozundan B, Saravanakumar K, Duran-Lara EF, Wang MH, Vaseeharan B (2019) Garlic clove extract assisted silver nano particle-antibacterial, antibiofilm, anti-helminthic, anti-inflammatory, anticancer and ecotoxicity assessment. *J Photochem Photobiol B.* 198111558. https://doi.org/10.1016/jphotobiol.2019.111558.

Wagner H, Breu W, Sendl A, Steinke B (1991) Comparative Studies of Extracts and Pure Compounds from Garlic, Wild Garlic and Onion on the Arachidonic Acid Metabolism and Platelet Aggregation. In *2nd International Symposium Pharmacy and Pharmacology and Clinical Applied Allium sativum*, March 7–10, Berlin, p. 15.

Wallock-Richards D, Doherty CJ, Doherty L, Clarke DJ, Place M, Govan JR, Compopiano DJ (2014) Garlic revisited: Antimicrobial activity of allicin containing garlic extracts against *Burkholderia cepacia* complex. *PLoS ONE.* 9: e112726.

Walton L, Herbold M, Lindegren CC (1936) Bactericidal effects of vapors from crushed garlic. *J Food Sci.* 1: 163–169.

Wang G, Yang Y, Wang C, Huang J, Wang X, Liu Y, Wang H (2020) Exploring the role and mechanisms of diallyl trisulfide and diallyl disulfide in chronic constriction-induced neuropathic pain in rats. *Korean J Pain*. 33(3): 216–225. https://doi.org/10.3344/kjp.2020.33.3.216.

Wang J, ZHang Xi, Lan H, Wang W (2017a) Effect of garlic supplement in the management of type 2 diabetes mellitus (T2DM): A meta-analysis of randomized controlled trials. *Food Nutr Res*. 61: 1377571. https://doi.org/10.1080/16546628.2017.1377571.

Wang L, Jiao H, Zhao J, Wang X, Sun S, Lin H (2017b) Allicin alleviates reticuloendotheliosis virus-induced immunosuppression via ERK/Mitogen activated protein kinase pathway in specific pathogen-free chickens. *Front Immunol*. 8: 1856. https://doi.org/10.3389/fimmu.2017.01856.

Wang TG, You WC, Handerson BE, Blot WJ (1985) A case control study of cancer in Shandong Province. *Natl Cancer Inst Monogr*. 69: 9–10.

Warber S (1999) Modes of action at target sites. In *Natural Products from Plants*. Kaufman PB, Cseke LJ, Warber S, Duke JA, Brielmann HL, Eds. CRC Press, New York, pp. 157–181.

Wargovich MJ, Goldberg MT (1985) Diallylsulfide: A naturally occurring thioether that inhibit carcinogen-induced nuclear damage to colon epithelial cells *in Vivo*. *Mutat Res*. 143: 127–129.

Wattenberg LW (1983) Inhibition of neoplasia by minor dietary constituents. *Cancer Res*. 43: 2448–2453.

WCRF/AICR (World Cancer Research Fund/American Institute for Cancer Research) (2007) *Food, Nutrtion, Physical Activity and the Prevention of Cancer: A Global Prespective*. AICR, Washington, DC.

Weber ND, Andersen DO, North JA, Murray BK, Lawson LD, Hughes BG (1992) In vitro virucidal effects of Allium sativum (garlic) extract and compounds. *Planta Med*. 58(5): 417–423. https://doi.org/10.1055/s-2006-961504.

Wei Z, Shan Y, Tao Li, Liu Y, Zhu Z, Liu Z, Wu Y, Chen W, WAng A, Lu Y (2017) Diallyl trisulfides, a natural histone deacetylase inhibitor attenuate HIF-1α synthesis and decreases breast cancer metastasis. *Mol Carcinog*. 56(10): 2317–2331. https://doi.org/10.1002/mc.22686.

Whanger PD (2004) Selenium and its relationship to cancer: An update dagger. *Br J Nutr*. 91: 11–28.

Wills E (1956) Enzyme inhibition by allicin, the active principle of garlic. *Biochem J*. 63: 514–520.

Wiseman M, Cannon G, Butrum R, Martin G, Higginbotham S, Heggie S, Jones C, Fletcher M (2007) *Food, Nutrition, Physical Activity and the Prevention of Cancer: A Global Perspective: A Project of World Cancer Research Fund International*. American Institute of Cancer Research, Washington, DC.

Wong JJ, Hsieh DPH (1976) Mutagenicity of Aflatoxins related to their metabolism and carcinogenic potential. *Proc Natl Acad Sci USA*. 73: 2241–2244.

Xiao D, Lew KL, Kin YA, Zeng Y, Hahm ER, Dhir R, Singh SV (2006) Diallyl trisulfide suppresses growth of PC-3 human prostate cancer Xenograft *in Vivo* in association with Bax and Bak induction. *Clin Cancer Res*. 12: 6836–6843.

Xiao D, Pinto JT, Soh JW, Deguchi A, Gundersen GG, Palazzo AF, Yoon JT, Shirin H, Weinstein IB (2003) Induction of apoptosis by the garlic-derived compound S-allyl mercaptocysteine (SAMC) is associated with microtubule depolymerization and C-Jun NH (2)-terminal kinase 1 activation. *Cancer Res*. 63(20): 6825–6837.

Xiao X, Chen B, Liu X, Liu P, Zheng G, Ye F, Tang H, Xie X (2014) Diallyl disulfide suppresses SRC/Ras/ERK signaling-mediated proliferation and metastasis in human breast cancer by up-regulating miR-34a. *PLoS ONE.* 9(11): 0112720. https://doi.org/10.1371/journal.pone.0112720.

Xu B, Monsarrat B, Gairin JE, Girbal-Neuhauser E (2004) Effect of ajoene, anatural antitumor small molecule, on human 20s proteasome activity *in Vitro* and in human leukemic HL60 cells. *Fundam Clin Pharmacol.* 18: 171–180.

Yamada Y, Azuma K (1977) Evalution of the *in Vitro* antifungal activity of allicin. *Antimicrob Agents Chemother.* 11: 743–749.

Yamaji K, Sarker KP, Abeyama K, Maruyama I (2004) Antiatherogenic effects of an egg yolk-enriched garlic supplement. *Int J Food Sci Nutr.* 55: 61–66.

Yamasaki T, Teel RW, Lau BHS (1991) Effect of allixin, a phytoallexin produced by garlic, on mutagenesis DNA binding and metabolism of aflatoxin B1. *Cancer Lett.* 59: 89–94.

Yki-Jarvienen H, Koivisto VA (1986) The natural course of insulin resistance in type 1 diabetes. *N Engl J Food Sci.* 315(4): 224–230 http://doi.org/10.1056INEIM198607243150404.

Yoshida S, Kasuga S, Hayashi N, Ushiroguchi T, Matsuura H, Nakagawa S (1987) Antifungal activity of ajoene derived from garlic. *Appl Environ Microbiol.* 53: 615–617.

You WC, Blot WJ, Chang YS (1988) Diet and high risk of stomach cancer in Shandong, China. *Cancer Res.* 48: 3518–3523.

You WC, Blot WJ, Chang YS, Ershow A, An Q, Yang JT, Hendersen BE, Fraumeni Jr JF, Wang TG (1989) *Allium* vegetable and reduced risk of stomach cancer. *J Natl Cancer Inst.* 81: 162–164.

Yung-Lin C, Raghu R, Lu K-H, Liu C-T, Lin S-H, Lai Y-S, Cheng W-C, Lin S-H, Sheen L-Y (2013) Autophagy therapeutic potential of garlic in human cancer therapy. *J Tradit Complement Med.* 3(3): 159–162.

Yuvashree M, Ganesh RN, Viswanathan P (2020) Potential application of nanomulsified garlic oil blend in mitigating the progression of type 2 diabetes-mediated nephropathy in Wistar rats. *3 Biotech.* 10(6): 1–15.

Zain Al-Abdeen SS, Abdullah IT, AI-Salihi SS (2013) The synergism effect of aqueous garlic extract and ciprofloxacin against some multiresestant bacteria. *J Microbiol Biotech Res.* 3: 136–142.

Zhai B, Zhang C, Sheng Y, Zhao C, He X, Xu W, Huang K, Luo Y (2018) Hypoglycemic and hypolipidemic effect of S-allyl-cysteine sulfoxide (alliin) in D10 mice. *Sci Rep.* 8(1): 1–7. https://doi.org/10.1038/s41598-018-21421-X.

Zhang X, Zhu Y, Duan W, Feng C, He X (2015) Allicin induces apoptosis of the MGC-803 human gastric carcinoma cell line through the p38 mitogen-activated protein kinase caspase-3 signaling pathway. *Mol Med Rep.* 11: 2755–2760.

Zhang Y, Xingping Liu, Runan J, Zhuang X, Zhang Xi, Zhiming Li (2020) Phytochemicals of garlic: Promising candidates for cancer therapy. *Biomed Pharmacol.* 123(2020): 109730. https://doi.org/10.1016/j.biopha.2019.109730.

Zhang ZD, Li Y, Jiao ZK, Xi Yi ZZ, Zhi JHZ (2008) Effect of local application of allicin via gastroscopy on cell proliferation and apoptosis of progressive gastric carcinoma. *Chin J Integrated Traditional Western Med.* 28(2): 108–110.

Zhang ZM, Yang XY, Deng SH Xu W, Gao HQ (2007) Anti-tumor effects of polybutylcyanoacrylate nanoparticles of diallyl trisulfide on orthotopic transplantation tumor model of hepatocellular carcinoma in BA2B/C nude mice. *Chin Med J.* 120: 1336–1342.

Zhao C, Deng W, Gage FH (2008) Mechanisms and functional implications of adult neurogenesis. *Cell.* 132(4): 645–660.

Zhen H, Fang F, Ye DY, Shu SN, Zhu YF, Dong YS, Nie XC, Li G (2006) Experimental study on the action of allitridin against human cytomegalovirus *in vitro* inhibitory effects on immediate early genes. *Antiviral Res.* 72: 68–74.

Zhenyne F, Jiay LV, Gao X, ZHeng H, Shi S, Xu X, Zhang B, Wu H, Song Qingqiao (2023) Effect of garlic supplementation on components of metabolic syndrome: A systematic review, meta-analysis and meta-regression of randomized controlled trails. *BMC Complement Med Ther.* 23: 260. http://doi.org/10.1186/S12906-023-04038-0.

Zhou W (2003) Acute lymphangitis treated by moxibustion with garlic in 118 cases. *J Trdit Clin Med.* 23: 198.

Zhou Xi, Qian H, Zhang D, Zeng L (2020) Garlic intake and the risk of colorectal cancer: A meta-analysis. *Medicine.* 99: 1. https://doi.org/10.1097/MD.0000000000018575.

Ziae S, Hantoshzadeh S, Rezasoltani P, Lamyian M (2001) The effect of garlic tablet on plasma lipids and platelet aggregation in nulliparous preeclampsia. *Eur Obstet Gynecol Report Biol.* 99: 201–206.

Ziamajidi N, Abbasalipourkabir R, Lotfi F, Goodarzi MT (2022) Aqueous garlic extract alleviates oxidative stress and inflammation in retinal tissue of rats with diabetes type 2. *J Adv Med Biomed Res.* 30(139): 138–145.

5 Garlic as Pesticide

Three lakh known species of plants on earth are attacked by 400,000 species of insects as well as by foraging animals. Of these, insect consume 10 to 15% of plant growth, prompting plants to evolve ingenious defensive strategies to ward off herbivorous predators (Schultz, 2002). Some plants adopt mechanical strategies, such as thorns, etc., while other plant groups develop a range of secondary metabolites. Secondary metabolites are specialized chemicals that can be easily dispensable, allowing the plant to survive and thrive. According to Pichersky and Gang (2000) 15 to 25% of each plant genome is associated with these secondary metabolites and 800 to 1500 differentially regulated genes are involved in a plant's response to a single insect. Plants are not static; they respond to attack with rapid, long-lasting and variable biochemical reactions, followed by physiological and developmental changes (Schultz, 2002).

The foliage of plants can release volatile secondary metabolites that repel insects/animals or prevent the growth of other plants. All parts of a plant, including roots and bulbs, stems, leaves, flowers, fruits and seeds, can release chemicals into the surroundings (Zeng, 2008). Plant-released toxic molecules that specifically target microorganisms are termed phytoalexins or phytoanticipins. Both are low molecular weight antimicrobial compounds that are either present in plants before microbial challenge or produced after infection from preexisting constituents (Van Etten et al., 1994). Allicin and other thiosulfinates are phytoanticipins, protecting *Allium sativum* (garlic) against plant pathogens (Curtis et al., 2004; Slusarenko, 2008). Garlic-derived polysulfides are approved for use as nematicides and insecticides (Sethi et al., 2014). Garlic protects the plant against various infections and is used as natural pesticide (Dancewicz and Gabrys, 2008). Its strong aroma can mask normal host-finding or feeding cues in insects and provides an olfactory camouflage (Perrin and Phillips, 1978). Magwenya et al. (2016) noted that garlic extracts have sufficient pesticidal activity in suppressing aphid infestation on cotton.

5.1 GARLIC AS INSECTICIDE

Effective pesticides have been developed from garlic extracts. Field trials carried out on mosquito breeding sites at Bombay (now Mumbai, India) have shown that garlic oil is very effective on several species of mosquitoes. Amonkar and Reeves (1970) studied the toxic effects of crude methanolic extract and garlic oil against third-stage larvae of *Culex peus*, *C. tarsalis*, *Aedes aegypti*, *A. trisoriatus*, *A. sirensis* and third- and fourth-stage larvae of highly insecticide-resistant strains of *A. nigromaculis*. They reported that the partially purified oil fraction was more toxic than the crude extract. Later, the larvicidal principles of garlic oil were isolated and identified as diallyl disulfide and diallyl trisulfide (Amonkar and Banerji, 1971). Allicin inhibited malaria infection by inhibiting cysteine protease, which processes the

DOI: 10.1201/9781003646211-5

circumsporozoite protein (CSP) of *Plasmodium sporozoites* required for the invasion of host cells (Coppi et al., 2006). These compounds were fatal to *Culex pipiens quinquefasciatus*. Nakagawa et al. (1980) reported that raw garlic juice at a dose 5 ml/kg body weight caused death of rats due to stomach injury. Gupta and Sharma (1993) observed that root-dip treatment of tomato seedlings in 25 ppm allicin for 5 mins was effective against the nematode *Meloidogyne incognita*. Allicin concentrations of 200 and 100 ppm, used as bare-root dips for 30 mins, killed 83% and 87% of *M. incognita* of tomato seedlings, respectively. Gareth et al. (2006) reported the insecticidal activity of garlic juice against two dipterian pests, *Delia radicum* (LC50 = 0.4%) and *Muscadomestica* (LC50 = 2.2%). They reported that garlic juice was effective against all the life stages of both insects.

The homopteran sucking insect *Lipaphis erysimi* (mustard aphid) causes severe damage to various crops. This pest not only affects plants but also transmits single-stranded RNA luteoviruses while feeding, which causes disease and damage in the crop. The mannose-binding garlic leaf lectin has been found to be a potent control agent of *L. erysimi* on one hand and symbionin-mediated luteovirus transmission on the other (Banerjee et al., 2004).

Gareth et al. (2006) noted that garlic juice is toxic to dipterian pests *Delia radicum* (LC_{50} = 0.04%) and *Musca domestica* (LC_{50} = 2.2%). Garlic is often grown among flowers or root vegetables as a companion plant to protect other plants from being attacked by pests (Sethi et al., 2014).

Plate-Rueda et al. (2017) reported the insecticidal activity of essential garlic oil and its compounds against the mealworm beetle *Tenbrio molitor* under laboratory condition. Garlic essential oil significantly killed and repelled the larval, pupal and adult stages of *T. molitor*. Garlic oil caused a dose-dependent toxic effect within 12 hours of application (Plate-Rueda et al., 2017). The larvae of *T. molitor* were the most susceptible, followed by the pupae and adults. The LC_{50} and LC_{90} values for larvae (0.77–1.36%), pupae (2.37–4.01%) and adults (2.03–4.73%) indicate that small quantities of garlic oil are toxic to different stags of *T. molitor*. Garlic oil application or by fumigation was also highly effective against stored product pests such as *S. oryzal, S. Zemais, Sitotroga cerealella* and *T. castaneum* (Huang et al., 2000; Kimbaris et al., 2009; Yang et al., 2009).

Diallyl sulfide and diallyl disulfide demonstrated toxic activity against different developmental stages of *T. molitor*. Diallyl disulfide have stronger contact toxicity in larvae (LC_{50} = 57.68 mg/ml), pupae (LC_{50} = 55.13 mg/ml) and adults (LC_{50} = 81.52 mg/ml) than diallyl sulfide in larvae (LC_{50} = 117.1 mg/ml), pupae (LC_{50} = 48.86 mg/ml) and adults (LC_{50} = 85.97 mg/ml). Garlic essential oil and its constituents, diallyl sulfide and diallyl disulfide, are highly toxic to *S. zeamais* and *T. castaneum* (Ho et al., 1996; Huang et al., 2000).

The rapid toxicity of garlic essential oil and their constituents in insects indicates neurotoxic action, as reported against *Delia radicum*, *Musca domestica*, *Cacopsylla chinensis* (Yang et al., 2012) and *Diaphorina citri*, with symptoms including hyperactivity, hyperextension of the legs and abdomen and a rapid knockdown effect (Mann et al., 2011). Acetylcholinesterase, an enzyme found at nerve synapses, is inhibited by diallyl disulfide, deallyl trisulfide and allicin (Bhatnagar et al., 1974).

In tropical areas, insect infestation—particularly from *Coleoptera* (beetles), *Lepidoptera* (moths) and *Isoptera* (termites)—is a major cause of damage and food loss. Methyl bromide is commonly used to kill all these infestations, but it has a negative impact on the stratospheric ozone layer. Its use is banned under the Montreal Protocol. Auger et al. (1999) suggested the use of garlic thiosulfinates and dimethyl and diallyl disulfides as fumigants. Diallyl trisulfide was found to be more potent than diallyl disulfide (Huang et al., 2000; Park et al., 2005). Both dimethyl and diallyl disulfide were effective at levels of Ca 0.001 M in inhibiting 90% of the growth of different species of fungi. Fungi showed more resistance to the disulfides compared to insects (Auger et al., 2004). Gautier et al. (2008) demonstrated that dimethyl disulfide targets calcium-activated potassium channels in insect pacemaker neurons, which regulate vital functions, such as heart wall muscle activity.

Garlic extracts are toxic across all life stages of insects belonging to the orders *Coleoptera*, *Lepidoptera*, *Heteroptera* and *Diptera*. Aqueous extracts of garlic inhibit egg hatching in mosquitoes (Jarial, 2001). Garlic extracts and steam distillates have toxic and antifeedant effects on stored product pests of the order *Coleoptera* (Chaim et al., 1999) and *Lepidoptera* (Gurusubramanian and Krishna, 1996), as well as repellent and toxic effects on *Homoptera*. According to Mandal and Choudhuri (1982), the toxic effect of garlic extract against the giant red bug (*Lohita grandis*) is due to its antibiotic action on symbiotic intestinal cholesterol, which is essential and cannot be synthesized by the insect itself.

Oparaeke et al. (2007) found that spraying of 5%, 10% and 20% garlic extract at different application intervals was effective in reducing insect populations and suggested that garlic extract is a viable alternative to synthetic pesticides for cowpea management in developing countries with limited resources. Samarasinghe et al. (2007) reported that methanolic garlic extract was effective in killing the cabbage moth *Plutella xylostella*. A 5.0% methanolic extract of garlic caused 67% mortality of first instar larvae within 48 hours.

The repellent and anti-feeding effects of garlic oil against sand flies were reported by Valerio and Maroli (2005). Topical application of garlic oil on the skin of volunteers provided 97% and 40% protection at 1% and 0.005% dilutions, respectively. In trials using an artificial membrane feeding apparatus, garlic oil showed a dose-dependent anti-feeding effect against *Phlebotomus papatasi* (sand fly), with 100% and 38% inhibition at 0.1% and 0.005% dilutions, respectively. A fresh garlic extract retained 83% of its original activity in killing mosquito larvae (LD_{50} = 0.1g/L) after 90 hours (Thomas and Callaghan, 1999). Garlic oil components repel mosquitoes (Gries et al., 2009).

5.2 GARLIC AS MOLLUSCICIDE

Singh and Singh (1993) reported that the water extract of garlic bulb is a potential source of molluscicide against *Lymnaea acuminata* and *Indoplanorbis exustus*. These snails are the intermediate hosts of the liver flukes *Fasciola hepatica* and *F. gigantica*, which cause 94% of fascioliasis in the buffalo population of northern India (Singh and Agarwal, 1981; Singh and Agarwal, 1983). Singh and Singh (1995) characterized allicin as the main molluscicidal component of garlic. The toxicity of allicin (LC50 = 3.64 mg/l) after 96 hours of exposure is 4.1, 4.2 and 2.5 times higher compared to the synthetic molluscicides phorate, carbaryl and formothion, respectively (Singh and Agarwal, 1983). There was a progressive increase in the toxicity (8.67-fold) of garlic bulb extracts against *Lymnaea acuminata* in the preharvest period (two to five months), while in the postharvest period (5 to 11 months), the increase was 5.5-fold (Singh and Singh, 1996a). The significant increase in the toxicity of preharvest garlic is due to the increase in alliin levels. An increase in molluscicidal activity of garlic in the postharvest period is due to some changes in the chemical composition of the garlic bulb (Lawson et al., 1991). Singh and Singh (1996b) demonstrated that treatment of snails with sublethal concentrations of allicin caused significant inhibition of acetylcholinesterase (AChE), lactic dehydrogenase (LDH) and alkaline phosphatase (ALP) activities in the nervous tissue of *Lymnaea acuminata*. Kinetic studies demonstrated that the inhibition of AChE by allicin was uncompetitive, whereas the inhibition of LDH and ALP was competitive (Singh and Singh, 1996b). The inhibition of these enzymes by allicin in the nervous tissue of the snail *L. acuminata* may be the cause of the molluscicidal action of garlic.

Singh and Singh (1997), Singh et al. (1998), Singh and Singh (2000a, 2000b), Singh and Singh (2001a, 2001b) and Tripathi and Singh (2001) reported that combinations of garlic bulb powder with *Cedrus deodara* oil, *Annona squamosa* seed, *Lawsonia inermis* seed, *Punica granatum* bark, *Canna indica* root powder, piperonyl butoxide and MGK-264 are very effective molluscicide. These combinations were effective in reducing the fecundity, hatchability and survival of young snails (Singh and Singh, 2004). Singh and Singh (2000b) have clearly demonstrated that these effects of garlic powder were due to the inhibition of certain enzymes and the alteration in brain biogenic amine levels in the nervous tissue of *L. acuminata*. Mantawy and Mahmoud (2002) observed that glucose and glycogen levels, as well as phenoloxidase activity in the snail *Biomphalaria alexandrina*, were significantly decreased after two and seven days of feeding on garlic, which ultimately reduced the fecundity of the snails. Spraying garlic powder (24 hours LC50 = 7.24%) alone and in combination with *Cedrus deodara* (24 hours LC50 = 4.18%) was toxic and effective in controlling the reproductive capacity of the giant African snail *Achatina fulica* (Rao and Singh, 2000, 2002; Rao et al., 2003). Allicin, the active molluscicidal component of garlic (Singh and Singh, 1995), is effective in killing snails, as well as making them sterile (Singh and Singh, 2000a). Tripathi et al. (2006), while studying the molluscicidal effect of cow urine, reported that a binary combination of cow urine plus garlic bulb powder is very effective in killing the vector snail *Lymnaea acuminata*. Kumari and Singh (2011), for the first time, reported that the larvae of fasciola could be killed within the body of a vector snail by phytotherapy of infected snails using sublethal concentrations of allicin. They also noted that allicin is highly toxic against radia and cercaria larvae by inhibiting acetylcholinesterase and cytochrome oxidase activity in the larvae (Kumari et al., 2016).

Picardal et al. (2018) found that the garlic acts as potential plant-based molluscicide. According to them, garlic can control the most invasive rice snail, *Pomecea canaliculata*. Garlic bulb extract (100%) showed molluscicidal activity comparable to that of the proven synthetic molluscicide niclosamide. Aged garlic extract treatment caused several behavioral changes in the snail *Pomacea canaliculata*, such as hyperactivity, twisting motion and the release of a white, sticky, saliva-like substance. This sticky substance was probably triggered upon the reaction of the snail's mucous membrane towards the polar compounds of *Allium sativum* (Picardal et al., 2018). The phytochemical allicin found in garlic is highly irritating in nature and affects the mucosal membrane. The Malacology Research Laboratory, Department of Zoology, D.D.U. Gorakhpur University, Gorakhpur, U.P., India, has been working on fasciolosis control in eastern Uttar Pradesh for four decades. The laboratory has studied the use of garlic formulations to control fasciolosis. Garlic extracts and allicin are used as molluscicides and cercaricides. This group of scientists has worked on different aspects of garlic as a molluscicide and cercaricide, both alone and in combination with cow urine and other formulations, for the effective control of fasciolosis (Kumari and Singh, 2011; Kumari et al., 2013; Kumari et al., 2015; Kumari et al., 2016; Kumar et al., 2021). Stored garlic bulb (11 months old) is more effective against *L. acuminata* and *I. exustus* than fresh garlic bulb (Singh and Singh, 1996a). Later, they reported that allicin coused uncompetitive inhibition of acetylcholinesterase in the nervous tissue of snails exposed to garlic extracts (Singh and Singh, 1996b). Reda et al. (2024) reported that garlic extracts exerted antiparasitic potential against gill monogeneans in both *in vitro* and *in vivo* conditions.

5.3 GARLIC AS ARACHNIDACIDE (MITES AND TICKS)

Fowl mites (*Ornithonyssus sylviarnm*) are external parasites that can reduce egg production and cause anemia and death in laying hens. When mite-infested hens were sprayed with water or 10% garlic juice for one to three weeks, a significant reduction in the number of mites was observed (Birrenkott et al., 2000). Maurer et al. (2009) found that a garlic juice preparation was effective *in vitro* against poultry red mites (*Dermanyssus gallinae*). In the United Kingdom, a garlic juice product has been approved for use against poultry red mites.

Asian bee mite *Tropilaelaps clareae* is a serious threat to beekeeping, affecting both developing brood and adult honey bees. It can cause bees to abscond from the hive. The natural host of the mite is the giant Asian honey bee, *Apis dorsata*, but *T. clareae* can readialy infest colonies of *A. mellifera*, the Western honey bee. Hosamani et al. (2007) reported that a single spray of 92% aqueous garlic extract resulted in significant mortality of *T. clareae* mites in the hive. A single spray of the garlic extract led to a significant increase in the untreated group. It did not affect the organoleptic quality of the honey or the strength of the bee colony.

5.4 GARLIC AS NEMATOCIDE

Nematodes inhabit the topsoil at levels of up to 1 billion per acre and cause immense economic losses, particularly to root crops. Garlic extract have been found to be repellent to the nematode *C. elegens*. Sukul et al. (1974) reported that under *in vitro* conditions, garlic extracts kill root-knot nematodes within 40 minutes. Aqueous/methanolic garlic extracts, garlic oil and synthetic diallyl disulfide are potent nematicides against *Aphelenchoides sacchari* and *Tylenchulus semipenetrans* (Nath et al., 1982). Aqueous and methanolic extracts completely killed *A. sacchari* and *T. semipenetrans* at relatively high concentrations of 0.5%, whereas garlic oil and synthetic diallyl disulfide were much toxic, killing *A. sacchari* at 8 and 25 ppm, respectively (Nath et al., 1982). Park et al. (2005) found that the garlic oil component diallyl trisulfide (LC_{50} = 2.79 µl/L) is more potent than diallyl disulfide (LC_{50} = .3706 µl/L) against the pine wood nematode *Bursaphelenchus xylophilus*. A five-minute immersion of tomato roots in 25 µg/L allicin inhibited juvenile penetration of the roots by 50% and was not phytotoxic. Allicin inhibited the hatching of the nematode *Meloidogyne incognita* at a concentration of 0.5 mg/ml and was toxic to juveniles at 2.5 µg/ml (Gupta and Sharma, 1993).

Garlic granules formulated from a combination of wood flour (55% w/w) and garlic juice (45% w/w), known to contain garlic oil polysulfides, were dissolved in water at concentrations of 25 or 12 mg/ml and caused mortality of the needle nematode *L. elongatus*, the potato cyst nematode *G. pallida*, the stubby root nematode *P. anemones* and the root-knot nematode *M. hapla* within four hours (Groom, 2004).

REFERENCES

Amonkar SV, Banerji A (1971) Isolation and characterization of larvicidal principle of garlic. *Science*. 174–1343.

Amonkar SV, Reeves EL (1970) Mosquito control with active principle of garlic (*Allium sativum*). *J Econ Entomol*. 63: 1172–1175.

Auger J, Arnault I, Diwo-Allain S, Ravier M, Molia F, Pettiti M (2004) Isecticidal and fungicidal potential of *Allium* substances as biofumigants. *Agroindustria*. 3: 5–8.

Auger J, Cadou XF, Thibout E (1999) *Allium* spp. thiosulfinates as substitute fumigants for methyl bromide. *Pestic Sci*. 55: 200–202.

Banerjee S, Hess D, Majumder P, Roy D, Das S (2004) The interactions of *Allium sativum* leaf agglutinin with a chaperon in group of unique receptor protein isolated from a bacterial endosymbiont of the mustard aphid. *J Biol Chem*. 279: 23782–23789.

Bhatnagar-Thomas PL, Pal AK (1974) Studies on the insecticidal activity of garlic oil 2. Mode of action of the oil as a pesticide in *Musca domestica* nebulo Fabr and *Trogoderma granarium* Everts. *J Food Sci Technol*. 11: 153–158.

Birrenkott GP, Brockenfelt GE, Greer JA, Owens MD (2000) Topical application of garlic reduces northern fowl mite infestation in laying hens. *Poultry Sci*. 79: 1575–1577.

Chaim WY, Huang Y, Chen SX, Ho SH (1999) Toxic and antifeedant effects of allyl disulfide on *Tribolium castaneum* (Coleoptera: Tenebrionidae) and *Sitophilus zeamais* (Coleoptera: Curculionidae). *J Econ Entomol*. 92: 239–245.

Coppi A, Cabinian M, Mirelman D, Sinnis P (2006) Antimalarial activity of allicin, a biologically active compound from garlic cloves. *Antimicrobial Agents Chemother*. 50: 1731–1737.

Curtis H, Noll U, Stormann J, Slusarenko A (2004) Broadspectrum activity of the volatile phytoanticipien allicin in extracts of garlic (*Allium sativum* L.) against plant pathogenic bacteria, fungi and oomycetes. *Physiol Mol Plant Pathol*. 65: 79–89.

Dancewicz K, Gabrys B (2008) Effect of extracts of garlic (*Allium sativum* L.), wornwood (*Artemisia absinthium* L.) and tansy (*Tanaceum vulgare* L.) on the behavior of the peach potato aphid (*Myzuspersical sulz.*) during the settling on plants. *Pestycydy/Pesticides*. 3–4: 93–99.

Gareth MP, Tamara SG, Andrew F (2006) Insecticidal activity of garlic juice in two dipterian pests. *Agric For Entomol*. 8: 1–6.

Gautier H, Auger J, Legros C, Lapied B (2008) Calcium-activated potassium channels in insect pacemaker neurons as unexpected target site for the noval fumigant dimethyl disulfide. *J Pharmacol Exp Ther.* 324: 149–159.

Gries RM, Gries GG, Khaskin G, Avelino N, Cambell C (2009) *Compounds, Compositions and Methods of Repelling Blood—Feeding Arthropods and Deterring Their Landing and Feeding.* U.S. Pat. Appl. US 2009/0069407 A1.

Groom MR (2004) *Ecospray Products as a Nematicide: A Report Submitted to the UK Pesticide Safety Directorate.* https://www.pan-europe.info/old/Archive/conferences/AnnualConf2006/MurreeGroomSession2.pdf.

Gupta R, Sharma NK (1993) A study of the nematicidal activity of allicin—an active principle in garlic, *Allium sativum* L., against root knot nematode, *Meloidogyne incognita* (Kojoid and White, 1919) Chitwood 1949. *Int J Pest Manag.* 39: 390–392.

Gurusubramanian G, Krishna SS (1996) The effect of exposing eggs of four cotton insect pests to volatiles of *Allium sativum* (Liliaceae). *Bull Entomol Res.* 86: 29–31.

Ho SH, Koh L, Ma Y, Huang Y, Sim KY (1996) The oil of garlic, *Allium sativum* L. (Amaryllidaceae), as a potential grain protectant against *Tribolium castaneum* (Herbst) and *Sitophilus zeamais* Mostsch. *Postharvest Biol Technol.* 9: 41–48.

Hosamani RK, Gulati R, Sharma SK, Kumar R (2007) Efficacy of some botanicals against ectoparasitic mite, *Tropilaelaps clareae* (Acari: Laelapidae) in *Apis mellifera* colonies. *Systemat Appl Acarol.* 12: 99–108.

Huang Y, Chen SX, Ho SH (2000) Bioactivities of methyl allyl disulfide from essential oil of garlic to two species of stored-products pests, *Sitophilus zeamais* (Coleoptera Curculionidae) and *Tribolium castaneum* (Coleoptera Tenebrionidae). *J Econ Entomol.* 93: 537–543.

Jarial MS (2001) Toxic effect of garlic extracts on the eggs of *Aedes aegypti* (Diptera: Culicidae): A scanning electron microscopic study. *J Med Entomol.* 38: 446–450.

Kimbaris AC, Kiouloos E, Koliopulos G, Polissiou MG, Michaelaks A (2009) Coactivity of sulfide ingredients; a new perspective of the larvicidal activity of garlic essential oil against mosquitoes. *Pest Manag Sci.* 65: 249–254.

Kumar P, Kumari S, Singh DK (2021) *In Vitro* activity of different combinations against *Fasciola gigantica. Curr Life Sci.* 2(3): 58–63.

Kumari S, Habib M, Kumar P, Singh VK, HuSAin SA, Singh DK (2016) Inhibition of acetylcholinesterase and cytochrome oxidase activity of *Fasciola gigantica* cercaria by phytoconstituents. *Acta Tropica.* 154: 19–24.

Kumari S, Kumar P, Singh VK, Singh DK (2013) *In Vitro* phytotherapy of vector snails by binary combinations of larvicidal active components in effective control of fasciolosis. *Rev Ins Med Trop.* 5(3): 303–308.

Kumari S, Kumar P, Singh VK, Singh DK (2015) Effect of phytocercaricide on certain enzyme activity in parasitic cercaria larva of *Fasciola gigantica. Eur J Biol Res.* 5(3): 52–57.

Kumari S, Singh DK (2011) Fascioliasis control: *In Vivo* and *in Vitro* phytotherapy of vector snail to kill fasciola larva. *J Parasitol Res.* 2011, article ID 240807. https//doi.org/10.1155/2011/240807.

Lawson LD, Wang ZJ, Hughes BG (1991) Identification and HPLC quantitation of the sulfinates in commercial garlic products. *Planta Med.* 57: 363–370.

Magwenya T, Svotwa E, Katsaruware RD (2016) Evaluating the efficacy of garlic (*Allium sativum*) as a biopesticide for controlling cotton aphid (*Aphis gossypii*). *Sci Agri.* 16(2): 54–60.

Mandal S, Choudhuri DK (1982) Cholesterol metabolism in Lohita grandis Gray (Hemiptera: Pyrrhocoridae: Insecta). Effect of corpora allatectomy and garlic extract. *Curr Sci.* 5: 367–369.

Mann RS, Rouseff RL, Smoot JM, Castle WS, Stelinski LL (2011) Sulfur volatiles from *Alliums spp.* Affect Asian citrus psyllid, *Diaphorina citri* Kuwayama (Hemiptera: Psyllidae), response to citrus volatiles. *Bull Entomol Res.* 101: 89–97.

Mantawy MM, Mahmoud AH (2002) Effect of *Allium cepa* and *Allium sativum* feeding on glucose, glycogen, protein bands profile and phenoloxidase activity in *Biomphalaria alexandrina. J Egypt Soc Parasitol.* 32: 271–283.

Maurer V, Perler E, Heckendorn F (2009) *In Vitro* efficacies of oils, Silicas and plant preperations against the poultry red mite *Dermanyssus gallinae. Exp Appl Acarol.* 48: 31–41.

Nakagawa S, Masamoto K, Sumiyoshi H, Kunihiro K, Fuwa T (1980) Effect of raw and extracted-aged garlic juice on growth of young rat and their organs after preoral administration. *J Toxicol Sci.* 5: 91–112.

Nath A, Sharma NK, Bhardwaj S, Thapa CD (1982) Nematicidal properties of garlic. *Nematologica.* 28: 253–255.

Oparaeke AM, Dike MC, Amatobi CI (2007) Effect of application of different concentrations and appropriate schedules of aqueous garlic (*Allium sativum* L.) bulb extracts against *Maruca vitrata* and *Clavigralla tomentosicollis* on cowpea, *Vigna unginiculata* (L.) Walp. Arch. Phytopath. *Plant Prot.* 40: 246–251.

Park IK, Park JY, Kim KH, Choi KS, Choi LH, Kin CS, Shin SC (2005) Nematicidal activity of plant essential oils and components from garlic (*Allium sativum*) and cinnamon (*Cinnamomum verum*) oils aginst the pine wood nematode (*Bursaphelenchus xylophilus*). *Nematology.* 7: 767–774.

Park SY, Cho SJ, Kwon HC, Lee KR, Rhee DK, Pyo S (2005) Caspase-independent cell death by allicin in human epithelial carcinoma cells imvolvement of PKA. *Cancer Lett.* 224: 123–132.

Perrin RM, Phillips ML (1978) Some effect on mixed cropping on the population dynamics of insect pest. *Entomol Appl.* 24: 385–393.

Picardal MT, Panlan KT, Castano PML, Pen LG, Abella KT, Picardal JP (2018) Molluscicidal activity of the aqueous extract of garlic (*Allium sativum* L.) bulb against golden apple snail (*Pamacea canaliculata* L.). *Int J Biosci.* 13(2): 75–87.

Pichersky E, Gang DR (2000) Genetics and biochemistry of secondary metabolites in plants: An evolutionary perspective. *Trends Plant Sci.* 5: 439–445.

Plate-Rueda A, Martinez LC, Santos MHD, Fernandes FL, Wilcken CF, Soares MA (2017) Insecticidal activity of garlic essential oil and their constituents against the mealworm beetle, *Tenebrio molitor* Linnaeus (Coleoptera: Tenebrionidae). *Sci Rep.* 7: 46406.

Rao IG, Singh DK (2000) Effect of single and binary combinations of plant derived molluscicides on reproduction and survival of the snail *Acatina fulica. Arch Environ Contaim Toxicol.* 39: 486–893.

Rao IG, Singh DK (2002) Toxic effect of single and binary treatments of synthetic and plant derived molluscicides against *Acatina fulica. J Appl Toxicol.* 22: 211–221.

Rao IG, Singh VK, Singh DK (2003) Effect of single and binary combination of plant derived molluscicides on different enzymes activities in the nervous tissue of *Achatina fulica. J Appl Toxicol.* 23: 19–22.

Reda R, Alshimaa AK, Mohamed E, Safaa IT, Enas AR (2024) Anti-parasitic activity of garlic (*Allium sativum*) and onion (*Allium cepa*) extracts against Dactylogyrus spp. (*Oreochromis nilotricus*): Hematology, immune response, hitopathological investigation and inflammatory cytokine genes of gills. *BMC Vet Res.* https://doi.org/10.1186//s12917-024-04/87-5.

Samarasinghe MKSRD, Chillar BS, Singh R (2007) Insecticidal properties of methanolic extract of *Allium sativum* L. and its fractions against *Plutella xylostella* (L.). *Pestic Res J.* 19: 145–148.

Schultz JC (2002) How plants fight dirty. *Nature.* 416: 267.

Sethi N, Kumar S, Dilbaghi N, Parle M, Pal M (2014) Garlic a pungent wonder from nature. *IRJP.* 5(7): 523–529.

Singh A, Singh DK (2001a) Molluscicidal activity of the clustard apple (*Annova squamosa* L.) alone and in combination with other plant derived molluscicides. *J Herbs Spices Med Plants.* 8: 29–29.

Singh A, Singh DK (2001b) Molluscicidal activity of *Lowsonia inermis* and its binary and tertiary combinations with other plant derived molluscicides. *Indian J Exp Biol.* 39: 263–268.

Singh A, Singh DK (2004) Effect of herbal molluscicides and their combination on the reproduction of the snail *Lymnaea acuminata. Arch Environ Contaim Toxicol.* 46: 470–477.

Singh DK, Agarwal RA (1983) *In Vivo* and *in Vitro* studies on synergism with anticholinesterase pesticide in the snail *Lymnaea acuminata. Arch Environ Contam Toxicol.* 12: 483–487.

Singh DK, Singh A (1993) Garlic (*Allium sativum*) a potent molluscicide. *Biol Agri Hortic.* 10: 175–178.

Singh K, Singh A, Singh DK (1998) The use of piperonyl butoxide and MGK-264 to improve the efficacy of some plant derived molluscicides. *Pest Sci.* 54: 145–149.

Singh K, Singh DK (1997) Molluscicidal activity of plant derived molluscicides. *J Herbs Spices Med Plants.* 5: 67–72.

Singh K, Singh DK (2000a) Effect of different combinations of MGK-264 or piperonyl butoxide with plant derived molluscicides on snail reproduction. *Environ Contaim Toxicol.* 38: 182–190.

Singh K, Singh DK (2000b) Toxicity of the snail *Lymnaea acuminate* of plant derived molluscicides in combination with synergists. *Pest Manag Sci.* 56: 889–898.

Singh O, Agarwal RA (1981) Toxicity of certain pesticides to two economic species of snail in northern India. *J Econ Entomol.* 74: 568–571.

Singh VK, Singh DK (1995) Characterization of allicin as a molluscicidal agent in *Allium sativum* (Garlic). *Biol Agric Hortic.* 12: 119–131.

Singh VK, Singh DK (1996a) Molluscicidal activity of pre and post harvest *Allium sativum* (Garlic). *Biol Agric Hortic.* 12: 311–318.

Singh VK, Singh DK (1996b) Enzyme inhibition of allicin the molluscicidal agent of *Allium sativum* L. (Garlic). *Phytother Res.* 10: 383–386.

Slusarenko AJ, Patel A, Portz D (2008) Control of plant diseases by natural products: Allicin from garlic as a case study. *Dur J Plant Pathol.* 121: 313–322.

Sukul NG, Das PK, De GC (1974) Nematicidal action of some edible crops. *Nematologica.* 20: 187–199.

Thomas CJ, Callaghan A (1999) The use of garlic (*Allium sativa*) and lemon peel (*Citrul limon*) extracts as *Culex pipiens* larvacides: Persistence and interaction with an organophosphate resistance mechanism. *Chemosphere.* 39: 2489–2496.

Tripathi R, Singh VK, Singh DK (2006) Freeze-derived powder of cow urine reduces the viability of the snail *Lymnaea acuminata. J Pest Sci.* 79: 143–148.

Tripathi SM, Singh DK (2001) Molluscicidal activity of *Punica grantum* and *Canna indica* combination with plant derived molluscicides against harmful snails. *Malas J Appl Biol.* 30: 25–31.

Valerio L, Maroli M (2005) Evaluation of repellent and anti-feeding effect of garlic oil (*Allium sativum*) against the bite of phlebotomine sandflies (Diptera: Psychodidae). *Ann Ist Super Sanita.* 41: 253–256.

Van Etten HD, Mansfield JW, Bailey JA, Farmer EE (1994) Two classes of plant antibiotics: Phytoalexins versus "phytoanticipins". *Plant Cell.* 6: 1191–1192.

Yang FL, Li XG, Zhu F, Lei CL (2009) Structure characterization of nano-dispersion particles loading garlic essential oil and its insecticidal activity against *Tribolium castaneum* (Herbst) (Coleoptera: Tenebrionidal). *J Agri Food Chem*. 57: 10156–10162.
Yang FL, Zhu F, Lie CL (2012) Insecticidal activities of garlic substances against moth, *Sitotroga cerealella* (Lepidoptera: Gelechiidae). *Insect Sci*. 19: 205–2012.
Zeng RS (2008) Allelopathy in Chinese ancient and modern agriculture. In *Allelopathy in Sustainable Agriculture and Forestry*. Zeng RS, Mallik A-V, Luo SM, Eds. Springer.

6 Adverse Effects of Garlic

Garlic is commonly used as an herbal alternative therapy in the treatment of various diseases. Adverse effects of garlic preperations have been reported with oral ingestion, inhalation and topical exposure, including bad breath, body odor, esophageal and abdominal pain and contact dermatitis. Despite its valuable medicinal properties, *Allium sativum* (garlic) may also act toxically when overdosed. Excessive garlic ingestion produces toxic effects (Rose et al., 1990; Fehri et al., 1991). The most apparent problem with using garlic in human medicine is its strong odor (Morbidoni et al., 2001). A clinically important issue is that some people are allergic to sulfur-based compounds. Several allergic reactions to garlic have been reported, namely, contact dermatitis, asthama, rhinitis, conjunctivitis, urticaria, anaphylaxis and angioedema (Gaddoni et al., 1994; Rauterberg, 1995; Kanerva et al., 1996; Armentia and Vega, 1997; Roberge et al., 1997; Morbidoni et al., 2001).

6.1 PER DAY SAFE GARLIC CONSUMPTION

Any product that is "natural" does not mean "SAFE". Researchers found that garlic homogenate caused liver damage after 21 days of use at a dose of 0.5 g/kg of rat body weight. They found that a dose 0.25 g/kg of rat body weight is safe, without harmful effects on the liver (Rana et al., 2006). Banerjee et al. (2002) noted that rats fed 2 g of fresh garlic per kg of body weight experienced 55% mortality after 15 days. The average consumption of garlic from all sources in the United States is 4.2 grams per day or more than one clove (Lucier, 2000). Allicin administered intraveneously to four rabbits (90.5% solution at 1 mL/min) caused pupil constriction, darkening of the blood, salivation and marked peripheral vasodilation. Death occurred at doses of 44 mg/kg (Block, 2010). Consumption of raw garlic—more than one or two cloves at a time—can irritate the lining of the digestive tract (Koch and Lawson, 1996). Consumption of garlic with food may reduce the risk of stomach irritation. *Alliums* are rich in polyphenolic compounds, which inhibit iron absorption by forming iron complexes. Tuntipopipat et al. (2008), in an *in vitro* study, found that alliums reduced iron availability in a dose-dependent manner.

6.2 *ALLIUM SATIVUM*–LINKED BOTULISM

Cases of botulism associated with the consumption of garlic-in-oil have been reported (Morse et al., 1990). Garlic is a bulb that can contain spores of the soil-dwelling organism *Clostridium botulinum*, which causes botulism. Garlic stored in a sealed container with olive oil provides good anaerobic conditions for *C. botulinum* to grow. The FDA recommends soaking garlic cloves for 24 hours in an acidic solution (such as citric, phosphoric acid or vinegar) to kill microorganisms. Drained garlic cloves, covered with edible oil and refrigenated, can then be used for up to 10 months.

DOI: 10.1201/9781003646211-6

6.3 CHOKING BY CONSUMPTION OF RAW GARLIC CLOVE

When a 60-year-old woman in Korea ate sliced raw fish served with garlic, she complained of severe, sustained chest pain. An endoscopic study revealed an intact garlic clove, 2.7 cm in size, lodged in the mid-esophagus. After removal of the clove, a 4 cm segment of whitish and bluish bullous necrotic change was observed. After three days, the patient's esophagus had fully healed. As it is well known that an intact, undamaged garlic clove has no medicinal value, it is suggested that it would be unwise to swallow it (Jabbari et al., 2005).

6.4 GARLIC ALLERGIES AND CONTACT DERMATITIS

Edelstein (1950), for the first time, noted contact dermatitis, allergic asthma and rhinitis from garlic preparations among cooks and garlic workers. Papageorgiou et al. (1983) and Hubbard Goldsmith (2005) found that garlic-sensitive patients had positive tests to diallyl disulfide, allyl propyl disulfide and allicin. Alvarez et al. (2003) recommended that diallyl disulfide be included in photopatch test series to identify photocontact allergy to diallyl disulfide. Diallyl disulfide penetrates most commercially available glove types (Moyle et al., 2004). Food allergies and anaphylactic reactions after consumption of raw garlic have been noticed in some cases (Valdivieso et al., 1994; Perez-Calderon et al., 2002; Moyle et al., 2004). In Taiwan, it has been reported that the enzyme alliinase found in the garlic bulb is a major allergen in a garlic-allergic group of patients. Several skin tests showed that this purified protein, alliinase, caused IgE-mediated hypersensitive responses in patients with garlic allergy (Kao et al., 2004). Although an entire garlic bulb produces little juice, it is a potent emetic. Garlic usually poses little safety issue; there are a few cases of topical garlic burns (Friedman et al., 2006) and anaphylaxis (Yin and Lin, 2007).

Burns associated with application of garlic, mostly for cure of various ailments, are a major concern in a number of medical reports. It is well known that raw garlic, containing allicin, is a potent skin irritant (Parish et al., 1987; Garty, 1993). Topical application of garlic caused blisterning lesions on children's skin (Rafaat and Leung, 2000). According to Lee and Lam (1991), the practice of direct application of fresh garlic onto the skin for treating infections should be discouraged. There is a significant body of literature on self-inflicted garlic burns (Friedman et al., 2006). In one of the papers, three cases—including burns on the upper arms of both female and male patients—were reported. Naturopathic physicians had treated all three individuals for localized pain. The treatment involved the application of garlic juice with a firm bandage on the site of pain. It was noted that all of the patients felt a burning sensation in the area but did not remove the bandage. After some time, erythema and small blisters developed in all three cases. Later, all the patients were treated with topical antibiotic ointments, and the burns healed within ten days.

Ekeowa-Anderson et al. (2007) decribe that a 40-year-old telephone engineer with a five-month history of worsening cheilitis, who had been taking several cloves of raw, chopped garlic for beneficial cardiovascular effects. Examination showed perioral and lip erythema and scaling with angular fissures. Bagga et al. (2008) studied an individual who sustained chemical burns of the oral mucosa caused by crushed garlic. The patient had placed crushed garlic cloves in the cheek overnight and developed a garlic burn injury. According to Jiang et al. (2008), the incidence of painful garlic burns is due to the misconception that allicin from fresh garlic might be a useful treatment for canker sores.

6.4.1 Effect of Garlic Consumption and Drug Therapy

It has been reported that garlic extracts inhibit the growth of *Mycobacterium avium*, which causes life-threatening pulmonary infections in immune-compromised AIDS patients (Deshpande et al., 1993). It has been also reported that garlic supplements significantly reduce the bioavailability of the anti-HIV protease inhibitor drug saquinavir. Patients receiving saquinavir as a protease inhibitor should avoid using garlic supplements (Piscitelli et al., 2002). Badol et al. (2007) noted that, in addition, garlic-derived thiosulfinates may interfere with thiol-containing drugs like Plavix®. Users of such drugs should avoid garlic consumption.

REFERENCES

Alvarez MS, Jacobs S, Jiang SB, Brancaccio RR, Soter NA, Cohen DE (2003). Photocontact allergy to diallyl disulfide. *Am J Contact Dermatol.* 14: 161–165.

Armentia A, Vega JM (1997) Can inhalation of garlic dust cause asthma? *Allergy.* 51: 137–138.

Badol P, David-Dufilho M, Auger J, Whiteheart SW, Rendu F (2007) Thiosulfinates modulate platelet activation by reaction with surface free sulfhydryls and internal thol-containing proteins. *Platelets.* 18: 481–490.

Bagga S, Thoma SBS, Bhat M (2008) Garlic burn as self-inflicted mucosal injury-a case report and review of literature. *Quintessence Int.* 39: 491–494.

Banerjee K, Maulik M, ManCahanda C, Dinda AK, Gupta SK, Maulik SK (2002) Dose-dependent induction of endogenous antioxidants administration of garlic. *Life Sci.* 70: 1509–1518.

Block E (2010) *Garlic and other Alliums (The Lore and the Science).* The Royal Society of Chemistry, Cambridge.

Deshpande RG, Khan MB, Bhat DA, Navalkar RG (1993) Inhibition of *Mycobacterium avium* complex isolates from AIDS patients by garlic (*Allium sativum*). *J Antimicrob Chemother.* 32: 623–626.

Edelstein AJ (1950) Dermatitis caused by garlic. *Arch Dermatol.* 61: 111.

Ekeowa-Anderson AL, Shergill B, Gold Smith P (2007) Allergic contact cheilitis to garlic. *Contact Dermatitis.* 56: 174–175.

Fehri B, Aiache JM, Korbi S, Monkni M, Ben MD, Memmi A, Hizaoui B, Bouker K (1991) Toxic effect induced by repeated administration of *Allium sativum* L. *J Pharm Belg.* 46: 363–374.

Friedman T, Shalom A, Westreich M (2006) Self-inflicted garlic burns: Our experience and litertature review. *Int J Dermatol.* 45(10): 1161–1163.

Gaddoni G, Selvi M, Resta F, Pasi F (1994) Allergic contact dermatitis to garlic in a cook. *Annl Ital Dermatol Clin Sper.* 48: 120–121.

Garty BZ (1993) Garlic burns. *Pediatrics.* 91: 658–659.

Hubbard G, Goldsmith P (2005) Garlic-fingered chefs. *Contact Dermatitis.* 52: 165–166.

Jabbari H, Argani A, Ghorbanihaghjo A, Mahdavi R (2005) Comparison between swallowing and chewing of garlic on levels of serum lipids, cyclosporine creatinine and lipid

peroxidation in renal transplant recipients. *Lipid Health Disease*. 4: 11. https://doi.org/10.1186/1476/511X-4-11.
Jiang XW, Hu J, Mian FI (2008) A new therapeutic candidate for oral aphthous ulcer: Allicin. *Med Hypotheses*. 71: 897–899.
Kanerva L, Estlander T, Jolanki R (1996) Occupational allergic contact dermatitis from species. *Contact Dermatitis*. 35: 157–162.
Kao SH, Hsu CH, Sci SN, Hor WT, Chang TWH, Chow LP (2004) Identification and immunologic characterization of an allergen alliin lyase from garlic (*Allium sativum*). *J Allergy Clin Immunol*. 113: 161–168.
Koch HP, Lawson LD, Eds. (1996) *Garlic: The Science and Terapeutic Applications of Allium Sativum L. and Related Species*. Williams and Wilkins, Baltimore, MD.
Lee TY, Lam TH (1991) Contact dermatitis due to topical treatment with garlic in Hong Kong. *Contact Dermatitis*. 24: 193–196.
Lucier G (2000) Garlic: Flavor of the ages. *Agric Outlook*, June/July, 7–10.
Morbidoni L, Arterburn JM, Young V, Mallins D, Mulrow C, Lawrence V (2001) Garlic: Its history and adverse effects. *J Herbal Pharmacother*. 1: 63–83.
Morse DL, Pickard LK, Guzewich JJ, Devine BD, Shayegani M (1990) Garlic-in-oil associated botulism: Episode leads to product modification. *Am J Public Health*. 80: 1372–1383.
Moyle M, Frowen K, Nixon R (2004) Use of gloves in protection from diallyl disulphide allergy. *Australasian J Dermatol*. 45: 223–225.
Papageorgiou C, Corbet JP, Menezes-Brandao F, Pecegueiro M, Benezra C (1983) Allergic contact dermatitis to garlic (*Allium sativum*). Identification of the allergens: The role of mono-di, and tri sulfides present in garlic. *Arch Dermatol Res*. 275: 229–234.
Parish RA, McIntire S, Heimbach DM (1987) Garlic burns: A naturopathic remedy gone awry. *Pediatr Emerg Care*. 3: 258–260.
Perez-Calderon R, Gonzalo-Garijo MA, Fernandez de S, Soria R (2002) Exercise-induced anaphylaxis to onion. *Allergy*. 57: 752–753.
Piscitelli SC, Burstein AH, Welden N, Gallicano KD, Falloon J (2002) The effect of garlic supplements on the pharmacokinetics of Saquinavir. *Clin Infect Dis*. 34: 234–238.
Rafaat M, Leung AK (2000) Garlic burns. *Pediatr Dermatol*. 17: 475–476.
Rana SV, Pal R, Vaiphei K, Singh K (2006) Garlic hepatotoxicity: Safe dose of garlic. *Tropical Gastroenterol*. 27: 26–30.
Rauterberg A (1995) Allergic contact dermatitis due to garlic in a cook. *Aktuelle Dermatol*. 21: 318–318.
Roberge RJ, Leckey R, Spence R, Krenzelok EJ (1997) Garlic burns of a breast. *Am J Med*. 15: 548.
Rose KD, Croissant PD, Parliament CF, Levin MB (1990) Spontaneous spinal epidural hematoma with associated platelet dysfunction from excessive garlic ingestion: A case report. *Neurosurgery*. 26: 880–882.
Tuntipopipat S, Zedar C, Siriprapa P, Charoenkiatkul S (2008) Inhibitory effects of species and herbs on iron availability. *Int J Food Sci Nutr*. 60: 43–55.
Valdivieso R, Subiza J, Verela-Losada S, Subiza JL Narganes MJ, Martinez-Cocera C, Cabrera M (1994) Bronchial asthma, rhinoconjunctivitis and contact dermatitis caused by onion. *J Allergy Clin Immunol*. 94: 928–930.
Yin J, Lin H (2007) Anaphylax's caused by younger garlic. *J Allergy Clin Immunol*. 119(1): 34.

Index

Note: Page numbers in *italics* indicate a figure and page numbers in **bold** indicate a table on the corresponding page.

H

I

L

M

N

O

P

R

S

T

U

V

W

For Product Safety Concerns and Information please contact our EU representative GPSR@taylorandfrancis.com Taylor & Francis Verlag GmbH, Kaufingerstraße 24, 80331 München, Germany